丝路书香工程项目

Chinese National Alcohols

BAIJIU AND HUANGJIU

BAOGUO SUN
Beijing Technology and Business University, China

TRANSLATED BY

JIHONG WU
Beijing Technology and Business University, China

LIANGLI (LUCY) YU
University of Maryland, USA

 **Chemical Industry Press Co., Ltd.**

内容简介

作为世界东方的文明古国，中国也是酒的故乡和酒文化的发源地。而白酒和黄酒都是中国独有的，是中华民族重要的非物质文化遗产，是中华文化最鲜明的符号。

本书以国酒（即白酒、黄酒）为题，用讲“故事”的表达形式，首次对中国白酒和黄酒进行了系统的归纳。在不长的篇幅内，以清晰、简明、通俗而严谨的语言，将中国白酒和黄酒的概念、特色、酿造的工艺、香型的区分与特点以及名酒的故事等进行了集中介绍，力争成为社会大众和海外消费者了解中国酒、研究中国酒、品味中国酒的一部百科全书。

图书在版编目（CIP）数据

国酒=Chinese National Alcohols：BAIJIU AND HUANGJIU：英文 / 孙宝国主编；吴继红，（美）俞良莉译. —北京：化学工业出版社，2021.7

ISBN 978-7-122-39557-3

Ⅰ.①国… Ⅱ.①孙… ②吴… ③俞… Ⅲ.①酒文化-中国-英文 Ⅳ.①TS971.22

中国版本图书馆CIP数据核字（2021）第137541号

责任编辑：赵玉清　　装帧设计：张　辉
责任校对：边　涛

出版发行：化学工业出版社（北京市东城区青年湖南街13号　邮政编码100011）
印　　装：北京宝隆世纪印刷有限公司
710mm×1000mm　1/16　印张17¼　字数320千字　2021年7月北京第1版第1次印刷

购书咨询：010-64518888　　售后服务：010-64518899
网　　址：http：//www.cip.com.cn
凡购买本书，如有缺损质量问题，本社销售中心负责调换。

定　　价：199.00元

Chinese National Alcohols

BAIJIU AND HUANGJIU

Chinese National Alcohols

BAIJIU AND HUANGJIU

BAOGUO SUN
Beijing Technology and Business University, China

TRANSLATED BY

JIHONG WU
Beijing Technology and Business University, China

LIANGLI (LUCY) YU
University of Maryland, USA

Published by

World Scientific Publishing Co. Pte. Ltd.
5 Toh Tuck Link, Singapore 596224
USA office: 27 Warren Street, Suite 401-402, Hackensack, NJ 07601
UK office: 57 Shelton Street, Covent Garden, London WC2H 9HE

and

Chemical Industry Press Co., Ltd.
No. 13 Qingnianhu South Street
Dongcheng District
Beijing 100011
P. R. China

Library of Congress Control Number: 2021934645

British Library Cataloguing-in-Publication Data
A catalogue record for this book is available from the British Library.

This edition is jointly published by World Scientific Publishing Co. Pte. Ltd. and Chemical Industry Press Co., Ltd.
B&R Book Program

CHINESE NATIONAL ALCOHOLS
Baijiu and Huangjiu

ISBN 978-981-123-356-2 (hardcover)
ISBN 978-981-123-357-9 (ebook for institutions)
ISBN 978-981-123-358-6 (ebook for individuals)

For any available supplementary material, please visit
https://www.worldscientific.com/worldscibooks/10.1142/12190#t=suppl

Typeset by Stallion Press
Email: enquiries@stallionpress.com

Printed in Singapore

Foreword

Standing out as a country with an ancient civilization in the oriental world, China is also the origin of alcoholic drinks and the cradle of alcoholic drink culture. It is well recognized that alcoholic drinks, connected closely with our daily life, are important consumer goods that satisfy both spiritual and material needs. Alcoholic drinks are often linked tightly with the culture of the individual country and region. Alcoholic drinks are critical components for human regional culture, arts, etiquette, customs, life, philosophy and aesthetics.

Exclusive to China, Baijiu and Huangjiu are regarded as the intangible cultural heritage of the Chinese nation and the most distinctive symbols of Chinese culture, and represent certain atmospheres, feelings or states of mind. With the desire for a better quality of life, the demand for better alcoholic drinks is increasing. However, most ordinary Chinese people feel the limitation of their knowledge about alcoholic drinks, and are eager to learn more. For instance, is Baijiu blending illegal or not? What are the 'Famous Eight' Baijiu brands that we usually talk about? What are the different flavor types of Baijiu? To date, not too many people really have the knowledge of Baijiu. The bias against Baijiu is common. How do we help consumers know more about Chinese alcoholic drinks in such an era when preaching is usually unacceptable?

Against the background of the Belt and Road Initiative and the 'Go Global' initiative of Chinese culture, it is important to not only inherit the skills and culture of Chinese alcoholic drinks but also advance the research on modern alcoholic drink culture and its link with Chinese culture, which helps international consumers understand Chinese alcohol and its cultural background. The top priority is to make sure that Chinese alcoholic drinks have their own unique Chinese names with Chinese cultural symbol(s), and not a name of a non-Chinese culture as in the past. For this purpose, Professor Baoguo Sun proposed first that 'Baijiu', the Chinese Pinyin of Chinese spirits, should be the general translation of Chinese spirits internationally. This is a significant step in the perception of Chinese alcoholic drinks worldwide.

To develop and expand the culture of Chinese alcoholic drinks, we should include both tradition and innovation, both history and the present, and both the present and the future. This book, under the title of *Chinese National Alcohols: Baijiu and Huangjiu*, introduces the concept, the brewing technique, the distinction of flavors, the characteristics of flavor types and the historic stories of well-known types of Baijiu and Huangjiu with illustrations and comprehensible but vivid language. Informative, interesting and readable, this book helps general consumers understand Baijiu, corrects the misinterpretation of some proper terms and promotes the culture of Chinese alcoholic drinks. The authors of this book are experts in researching and manufacturing Baijiu and Huangjiu, and the book contains their latest research findings and thoughts.

The popularization of the knowledge of alcoholic drinks is an important task of the China Alcoholic Drinks Association (CADA). I hope that this book, apart from publicizing, spreading and developing the correct cognition of Chinese alcoholic drinks, could work as a messenger who brings the culture of Chinese alcoholic drinks to the whole world and a witness who helps Chinese alcoholic drinks become a major player in the world spirits market.

Yancai Wang
Director of the China Alcoholic Drinks Association
19 November 2018

Preface

'Drinking alcohol is a joy for the common while a courtesy for the gentle.' Since ancient times, alcohol has been used to worship Heaven, gods, ancestors, and to mourn the dead because it is sacred. Alcohol was a great tribute to present to the emperor in ancient times. According to the tales, Di Yi made and presented alcohol to Da Yu, the founder of the Xia Dynasty, the oldest Chinese Dynasty on record. Today's Gujinggong Baijiu comes from the story going back over 1800 years from the late Eastern Han Dynasty. Alcohol can be used to treat illnesses based on the theories of traditional Chinese medicine that 'Alcohol is also medicine' and 'Alcohol is an introductory medium added to enhance the efficacy of traditional Chinese medicine'. Nowadays, alcohols are must-have drinks during celebrations, banquets and daily life. The preparation of alcoholic drinks has developed for over 9000 years in China, which can be considered the origin of alcohol. The original alcohols in China were made from rice, hawthorn berries, grapes, honey, etc. As time has gone by, Baijiu and Huangjiu, exclusive in China, have become the mainstream drinks among the alcoholic beverages in China. Only produced in China, Baijiu and Huangjiu became the 'National Alcohols' of China because of their long history and abundant cultural components. With a history over 7000 years long, Huangjiu is one of the three ancient alcoholic beverages in the world along with grape wine and beer. Baijiu, one of the earliest distilled spirits, takes its place in the top six distilled spirits in the world by enjoying a history of over 2000 years, with the others being Whisky, Brandy, Vodka, Gin and Rum. During Baijiu and Huangjiu processing, grains are the raw materials, whereas Jiuqu is the sacchariferous and fermentative agent, which ensures that they are rich in beneficial functional components and moderate drinking does good to human health.

Although the culture of Chinese alcohols is rich and profound, people know Baijiu and Huangjiu less than Whiskey, Brandy, Vodka and red wine. There is still a long way to go in the pursuit of making Baijiu and Huangjiu international alcoholic beverages, as the major consumers are Chinese as of now. It is necessary to realize modernization of production and market internationalization in the development and expansion of Baijiu and Huangjiu. The popularization of the knowledge about Baijiu and Huangjiu and the enhancement to cultural confidence are in high demand.

In recent years, we have been advocating drinking the proper amount of Chinese national alcohols, at the right time and using good manners, and making an effort to spread the transliteration of Baijiu and Huangjiu in foreign languages: The translation of Chinese alcoholic drinks and Chinese yellow drinks (rice wine as it was called in the past) should be 'Baijiu' and 'Huangjiu', respectively. The idea that 'both flavor- and health-oriented strategy is what we should insist upon for future Baijiu and Huangjiu research and development' has been widely accepted by the public.

Written using our research findings, this book contains nine parts, including concept, history, culture, celebrities, famous brands, etc., in order to provide readers with a comprehensive and objective overview of Chinese Baijiu and Huangjiu.

This book has been edited and prepared by a team of experts led by Professor Baoguo Sun, an academician of the Chinese Academy of Engineering and the President of Beijing Technology and Business University (BTBU). The other experts involved in the book's preparation are Dr. Jihong Wu, Miss Hehe Li, and Dr. Ning Zhang from Key Laboratory of Brewing Molecular Engineering of China Light

Industry (KLBMECLI) at BTBU, Professor Jian Mao at Jiangnan University, and Professor Mingquan Huang, Dr. Xiaotao Sun, Dr. Jinyuan Sun, and Professor Fuping Zheng from KLBMECLI at BTBU. Due to the limitation of knowledge and information, omissions and errors may exist in this book, and your comments and suggestions are highly appreciated.

Baoguo Sun, PhD, Professor
Academician of the Chinese Academy of Engineering,
President of BTBU, Director of Key Laboratory of
Brewing Molecular Engineering of China Light Industry,
Beijing, China
September 2018

Contents

Acknowledgments

The development and preparation of *Chinese National Alcohols: Baijiu and Huangjiu* was facilitated by a number of dedicated people at World Scientific Publishing Co. Pte. Ltd., Chemical Industry Press, Beijing Technology and Business University (BTBU), and Jiangnan University. We would like to thank all of them, with special mentions for Yuqing Zhao and Gang Wu of Chemical Industry Press and Max Phua, Lixi Dong, Ling Xiao and Steven Shi Hongbing of World Scientific. Without them, our dream of this book would not have come true. It has been a great pleasure and fruitful experience to work with them in transforming our manuscript into a very attractive printed book.

The material in this book has mainly been derived from our outstanding colleagues, and we appreciate their contributions. They include Dr. Jihong Wu, Miss Hehe Li, and Dr. Ning Zhang from KLBMECLI at BTBU, Professor Jian Mao at Jiangnan University, and Professor Mingquan Huang, Dr. Xiaotao Sun, Dr. Jinyuan Sun, and Professor Fuping Zheng from KLBMECLI at BTBU. Additionally, we would register our thanks to our excellent students for the data support provided by their research works.

We would like to thank China Alcoholic Drinks Association (CADA), Professor Liangli (Lucy) Yu at University of Maryland, and

Professor Lihua Gao and Professor Xiuting Li at BTBU for their support, and we have attempted to acknowledge the help we have received. We would also like to express our appreciation to the continued support of many companies in these years, such as Luzhoulaojiao Distillery Co., Ltd., Anhui Gujing Group Co., Ltd., Shandong Bandaojing Co., Ltd., Jiangsu Yanghe Distillery Co., Ltd., Shandong Jingzhi Liquor Co., Ltd., Hubei Jinpai Co., Ltd., Zhejiang Guyuelongshan Shaoxing Huangjiu Co., Ltd., Qinghai Huzhu Barley Wine Co., Ltd., Beijing Weishiyuan Food Technology Co., Ltd., Hebei Hengshui Laobaigan Liquor Co., Ltd., Chengde Qianlongzui Liquor Co., Ltd., Qingdao Langyatai Group Co., Ltd., Sichuan Gulin Langjiu Distillery Co., Ltd., Jiugui Liquor Co., Ltd., Beijing Shunxin Agriculture Co., Ltd., Sichuan Tuopai Shede Spirits Co., Ltd., and so on. It would be quite impossible for us to express our appreciation to everyone concerned for their collaboration in producing this book, but we would like to extend our gratitude. In particular, we would like to thank several professional societies in which we have published some of the material in this book previously. They are the American Chemical Society (ACS), Royal Society of Chemistry (RSC), Beijing Academy of Food Sciences, and Chinese Institute of Food Science and Technology (CIFST) for their conferences, proceedings, and journals, including *Journal of Agricultural and Food Chemistry*, *RSC Advances*, *Food Science*, and *Journal of Chinese Institute of Food Science and Technology*.

We would also like to acknowledge the support of many funding agencies in the past many years, such as the National Natural Science Foundation of China (No. 31830069) and the Ministry of Science and Technology of China.

Baoguo Sun, PhD, Professor,
Academician of the Chinese Academy of Engineering,
President of BTBU, Director of Key Laboratory of
Brewing Molecular Engineering of China Light Industry,
Beijing, China

Abbreviations

CADA — China Alcoholic Drinks Association.

HZ-HS — operation of steaming raw materials and fermented grains together.

HZ-XCA — means that after mixing the last Jiupei and crushed new grain materials proportionally, the mixtures are steamed simultaneously in the steamer barrel named Zengtong.

LWZ — which is a traditional process in the production of many flavor types of Baijiu in China. The marrow of this method is to steam and blend fermented grains five times with new grain materials. Under normal conditions, there are four layers of fermented grains in the cellar.

NAAC — National Alcohols Appraisal Conference.

NTCAS — National Technical Committee for Alcohol-making Standardization.

QZ-ECQ — means steaming grains and Jiupei separately and secondary clarification.

QZ-QCA — means that the raw materials and auxiliary materials are steamed separately, then mixed proportionally, and then the starter is added to the material mixture for the first fermentation.

QZ-QS — means steaming raw grains and Jiupei separately.

QZ-QS-SCQ — means steaming grains and Jiupei separately and quartic clarification.

QZ-XCA — refers to the raw materials steamed separately, and then mixed with the remaining fermented grains from the former fermentation cycle before adding Fuqu starter to continue fermentation.

RB — raw Baijiu.

XCA-PL — means that adding a certain amount of auxiliary material bran shell to the original fermented grains, and then mixing evenly and cooking.

XZ-HZ — that is to say, the fermented grains and raw grain powder are mixed in proportion, then steamed together.

XZ-HZ-HS — that is to say, the fermented grains and raw grain powder are mixed in proportion; the fresh Baijiu and the grain are steamed at the same time. After steaming, the fermented grains are spread to cool, sprinkled with Jiuqu and put into the cellar for XCA fermentation. Because distillers' grains should be used continuously, it is also called 'XZ fermentation'. The fermented grains (mother grains) can be continuously recycled for several years, and will never be lost, so it is called 'ten thousand years distillers' grains'.

Chapter 1
Concepts

1. Alcoholic Drinks

Alcoholic drinks, such as Baijiu, Brandy, Whiskey, Vodka, Rum, Gin, Huangjiu, beer and grape wine, refer to alcoholic beverages made from fermentation of grains, fruits, sugarcanes and honey rich in starches and/or sugar. Baijiu, Brandy, Whiskey, Vodka, Rum and Gin, as shown in Figure 1.1, are distilled spirits, while Huangjiu, beer and grape wine involve no distillation.

Baijiu is a type of distilled alcoholic drink and is made from grain fermentation induced by Daqu, Xiaoqu and Fuqu as possible saccharifying and fermenting agents through solid saccharification, fermentation, distillation in barrels (named Zengtong), long-time storage in porcelain jars and blending processes.

Brandy is the signature liquor from France. It is a kind of distilled spirit made from grapes or other fruits after fermentation by yeast, distillation, storage in oak barrels and mixing.

Whiskey is the signature alcohol from the United Kingdom, and is a distilled liquor made from barley and other cereals after fermentation by yeast, distillation, storage in oak barrels and mixing.

Vodka is the signature alcohol from Russia and Finland, and is made from cereals and potatoes. Vodka is a type of distilled liquor. The original liquid of vodka that comes from fermentation

Baijiu Brandy Whiskey Vodka Rum Gin

Figure 1.1. Typical representative products of the six distilled alcoholic drinks in the world.

by yeast and rectification is filtrated slowly by active carbon from birches to remove the fusel oil, aldehydes, acids, esters and other minor components.

Rum is a distilled liquor from Cuba made from sugarcane molasses or sugarcane juice through fermentation by yeast, distillation, storage in oak barrels and mixing.

Gin is a low-alcohol distilled liquor made from grains, which is the signature alcohol from the Netherlands. The base of it is made by fermentation and distillation before addition of junipers and other aromatic plants, and then dipping, distillation and mixing.

Huangjiu, commonly called 'liquid cake', is a fermented beverage made from rice and millets after addition of Jiuqu and yeast as sacchariferous and fermentative agents.

Beer, commonly called 'liquid bread', is a mild alcoholic beverage with carbon dioxide which is made from barley, hops and water after fermentation by yeast.

Grape wine is an alcoholic beverage made from fresh grapes or grape juice after fermentation by yeast. Grape wine is further classified into red wine, white wine and rose wine according to color, and into dry wine, semi-dry wine, semi-sweet wine and sweet wine such as dry red wine and dry white wine according to their sugar contents.

Exclusive in China, rice spirits, also called 'Jiuniang', 'Laozao', 'Chinese sweet spirits' or 'Li' in ancient times, are made from sweet rice after fermentation by Jiuqu. With a low alcohol content but a strong delayed effect, rice spirits are nutritious and can also be used in foods. The best-known food with rice spirits is 'Laozao Tangyuan' (glutinous rice balls with/without fillings in rice spirit soup).

As an alcoholic beverage, liquor has a remarkable association with the development of human society, culture and history in the last thousands of years. The historical stories about alcohols, such as 'Banquet at Hongmen (a dinner at Hongmen where treachery was planned)' and 'Deprivation of the military authority of generals with cups of spirits', are well known in China. In the poetry of the Tang and Song Dynasties, two shining pearls in the history of Chinese literature, the character 'alcohol' appears 5814 times in *Tang Poems* and 4892 times in *Song Poems*. Many lines spread across the ages are written after alcohol drinking such as 'Everything appears in its way, alcohols bring my poems' from *Response to Yuanming Tao's Drinking Alcohols* by Shi Su and 'In plucking chrysanthemums beneath the east hedge, I vacantly see the southern mountains afar' from *Drinking* by Yuanming Tao. The lines 'Gold, wood, water, fire and earth, and Bancheng Shaoguo Baijiu' from Emperor Qianlong and Minister Xiaolan Ji are well-known couplets quoted after drinking.

Spirits and culture are always associated with each other. In the new era, the wonderful Chinese alcohol culture should be recognized and enjoyed together with Chinese Baijiu and Huangjiu.

2. Baijiu

Baijiu, the national alcohol of China, is one of the unique distilled spirits of China, which has a long history of more than 2000 years.

Baijiu is normally prepared from grains with Daqu, Xiaoqu and Fuqu as the common saccharifying and fermenting starters, which are mixed with grains to saccharify and ferment simultaneously. The fermented mixture is distilled under solid-state conditions using special equipment named Zengtong to produce fresh Baijiu. The fresh distillates will be aged in porcelain jars to obtain the desired flavors. The final commercial products are blends of aged distillates, fresh distillate and water to obtain different formulations of spirit drinks.

The grains, such as sorghum, wheat, rice, glutinous rice, corn, millet, barley, buckwheat and highland barley, are commonly used as the starting materials for Baijiu production.

Daqu is made from wheat, pea, barley, highland barley and other raw materials through natural inoculation and cultivation of microorganisms in the environment, and contains a variety of microorganisms, enzymes and chemical substances. The size of Daqu is larger than that of Xiaoqu. The primary microorganisms in Daqu are mold, yeast, bacteria and actinomycetes.

Xiaoqu is made from rice flour, rice bran and other raw materials by inoculation and cultivation of parent Qu and appears in block, pellet or non-pelleted forms. The microorganisms in Xiaoqu primarily include mold and yeast.

Fuqu is a kind of non-pelleted Qu which is made from bran as the starting material and inoculated with pure mold.

Baijiu solid-state fermentation is usually carried out in pits or fermentation vats. The fermentation vat, kept underground, is called the Digang.

There are many varieties of Baijiu. To date, there have been 12 representative flavor types of Baijiu, such as the Strong (Nong), Light (Mild), Sauce (Jiang), Mixed (Jian or Nongjiang), Rice (Mi), Feng, Te, Dong, Chi, Sesame, Fuyu and Laobaigan flavor types. The representative Baijiu products of 12 flavor types are shown in Figure 1.2.

The flavor types of Baijiu may vary greatly, which cannot be completely included by the 12 representative flavor types. A new flavor of Baijiu is constantly under development. It should also be noted that the flavor may differ for the Baijiu products of the same flavor type. For example, Luzhou Laojiao Baijiu, Wuliangye Baijiu, Gujinggong Baijiu and Bancheng Shaoguo Baijiu all belong to the same flavor type, the Strong flavor, but their flavors are significantly different compared to each other. Additionally, Jingzhi Teniang Baijiu and Guojing Baijiu are both Sesame flavor types, but now they are called Zhi flavor and Guojing flavor, respectively. The future development

Figure 1.2. The typical products of 14 representative flavor types of Baijiu.

of the Baijiu flavor type should be flourishing in harmony and diversity.

Baijiu, together with Brandy, Whiskey, Vodka, Gin and Rum, are the six distilled spirits in the world. Baijiu is also the one with the longest history, the longest fermentation period, the most complex technology, the largest production, the lowest degree of internationalization and is produced in the least number of countries (only in China).

3. Huangjiu

Huangjiu, a unique non-distilled alcoholic drink in China, is the national alcohol of China with a history of more than 7000 years.

Distinguished in the world's liquor-making history, Huangjiu is one of the three ancient alcohols along with grape wine and beer. Mild but mellow and full-bodied, Huangjiu is the best embodiment of simplicity of Chinese culture and of the spirit of toughness integrated with softness.

According to the national standard (GB/T 13662-2008), Huangjiu is defined as a fermented beverage made from rice and millets after addition of Jiuqu and yeast as sacchariferous and fermentative agents. As the saying goes, rice is the flesh of Huangjiu, Qu is its bone and water is its blood. Huangjiu, a beverage unique to China, comes from unique starting materials and preparation techniques.

Huangjiu is classified according to product style, sugar content, place of origin, starting materials and Jiuqu, and manufacturing technique.

According to the product style, Huangjiu is classified into three categories. The Traditional type Huangjiu tastes refreshing and is made from rice, millet, corn and wheat after steaming, addition of Jiuqu, saccharification, fermentation, pressing, filtrating, boiling (sterilization), storage and blending. The Light type Huangjiu tastes refreshing and is made from rice, millets, corn and wheat after addition of Jiuqu (or enzyme and yeast) as the sacchariferous and fermentative agent after steaming, saccharification, fermentation, pressing, filtrating, boiling (sterilization), storage and blending. The Special type Huangjiu is of a special flavor but still sticks to the traditional style because of the change of starting and subsidiary materials and processing technique.

According to the sugar content, Huangjiu is classified into four groups. They are dry Huangjiu with a total sugar content ≤15 g/L; semi-dry Huangjiu with a 15 g/L < total sugar content ≤40 g/L; semi-sweet Huangjiu with a 40 g/L < total sugar content ≤100 g/L; and sweet Huangjiu with a total sugar content >100 g/L.

According to the origin of production, it is classified into Shaoxing Huangjiu, Dai County Huangjiu, Fang County Huangjiu, Jimo Huangjiu, Lanling Huangjiu and Longyan Chengang Huangjiu as shown in Figure 1.3.

Figure 1.3. Representative Huangjiu in China according to the origin of production.

4. Daqu Baijiu

Daqu Baijiu, as shown in Figure 1.4, is a product brewed with Daqu as the saccharifying and fermenting starter. According to the shape and size of the starter, it can be divided into Daqu and Xiaoqu, out of which Daqu is the most representative form of the traditional starter during Baijiu processing. The size of Daqu is relatively large, generally in the shape of brick, and it is crushed into Daqu powder before use.

The typical production process for Daqu involves several steps. First, the staring ingredients are wetted with water, followed by heaping up, grinding and mixing with water. Then, Daqu is shaped as bricks in the mold chamber and is cultured in the Qu room. After being constantly flipped over, Daqu becomes dry and is kept in the storehouse. Finally, the end product of Daqu is obtained. The different temperatures, humidity and heaping-up during Daqu production in different regions and distilleries have resulted in different performances of the low-temperature, medium-temperature and high-temperature Daqu, for possible production of different flavor types. The Light flavor type Baijiu production generally uses low-temperature Daqu, whereas the Strong flavor type Baijiu generally uses medium-temperature Daqu. The high-temperature Daqu is generally used in the production of the Sauce flavor type Baijiu. In the production of the Sesame flavor type Baijiu, high-temperature Daqu and Fuqu are often used.

Figure 1.4. Representative Daqu Baijiu products from China.

With Daqu as the saccharifying and fermenting starter, Qingcha (QCA, only using the starting grains as the ingredient) or Xucha (XCA, using the starting grains with the distillers' grains, also called Jiuzao, as the ingredients) as the brewing technique, Zengtong as the distillation device, porcelain jar or stainless steel as the storage container and a combination of manual and computer blending art, Daqu Baijiu is endowed with the unique flavor of Strong, Light, Sauce and other flavor types.

Fen Baijiu, Moutai Baijiu, Luzhou Laojiao Baijiu, Wuliangye Baijiu, Gujinggong Baijiu, Yanghe Baijiu, Laobaigan Baijiu, Erguotou Baijiu and Lang Baijiu are all Daqu Baijiu.

5. Xiaoqu Baijiu

Xiaoqu Baijiu, as shown in Figure 1.5, is brewed with Xiaoqu as the saccharifying and fermenting starter. Xiaoqu is made of rice flour or

Figure 1.5. Representative Xiaoqu Baijiu products from China.

rice bran, sometimes supplemented with a small amount of Chinese herbal medicine or polygonum hydropiper powder, with a small amount of white earth (Guanyin soil) as the base, inoculated with mother Qu, shaped using the appropriate amount of water and cultured under controlled temperature and humidity conditions. It is mainly used to produce the Rice, Light and Chi flavor types of Baijiu. The microorganisms in Xiaoqu mainly include mold and yeast. Compared with Daqu, Xiaoqu has a smaller morphology, fewer microorganisms, less dosage to be used, a shorter fermentation period and greater Baijiu yield. Xiaoqu Baijiu is regional, mainly produced in the South and Southwest of China.

The typical production process of Xiaoqu generally involves the following: soaking the ingredient in water and steaming, inoculation, cultivation in the house and moving out of the house to obtain the final dry Xiaoqu. During the brewing process of Xiaoqu Baijiu, the required amount of Xiaoqu ranges from 0.5% to 1%. The primary reason is that the microbial culture enlarges in the stage of 'cultivating bacteria' during the production process. At present, Xiaoqu has many kinds, which can be divided into medicinal starter and non-medicinal starter depending on whether one is adding Chinese herbal medicine or not. It can also be classified into common Baijiu starter and sweet Baijiu starter based on the different utilization purposes. Additionally,

Xiaoqu can be divided into grain starter with whole rice flour and bran starter with a small amount of rice flour or whole rice bran according to the primary ingredients, and pancake-like, pellet-like and non-pelleted starter by their appearance. The representative Xiaoqus include Sichuan medicine-free bran starter, Qionglai rice starter, Xiamen Baiqu, Guilin medicine Xiaoqu and Guangdong pancake-like starter.

Xiaoqu Baijiu can be divided into three categories according to its production technology, Jiuqu and starting ingredients. The first category is produced using rice as the primary ingredient, Xiaoqu as the saccharifying and fermenting starter and the ingredients are saccharified and fermented under the liquid state before liquid distilling, which is the production process of the Chi flavor type of Baijiu. The second category of Baijiu uses rice as the ingredient, Xiaoqu as the saccharifying and fermenting agent, solid-state microorganism cultivation, liquid fermentation and liquid distillation, which is the manufacturing technology of the Rice flavor type of Baijiu. The third is made by solid-state fermentation with sorghum, corn, wheat and other grains as the starting ingredients, Xiaoqu as the saccharifying and fermenting starter and is distilled in solid-state, which is the production process of Xiaoqu Baijiu in Sichuan Province. Xiaoqu Baijiu is a good base for making liqueur, which has mellow, soft, pure and sweet, and pure body characteristics.

Maopu tartary buckwheat Baijiu, Guilin Sanhua Baijiu, Guangdong Yubingshao Baijiu, Changle Baijiu and Chongqing Jiangxiaobai Baijiu all belong to Xiaoqu Baijiu.

6. Fuqu Baijiu

Fuqu Baijiu, shown in Figure 1.6, is brewed with purebred Fuqu and yeast as the saccharifying and fermenting starters. Fuqu uses bran as the carrier. After being steamed, sterilized and spread, the carrier is inoculated with pure microorganism strains, and cultured under controlled temperature and humidity. The primary microorganism in the Fuqu is mold. Fuqu mainly plays the role in saccharification, and is mixed with yeast (purebred cultured yeast) for alcoholic fermentation during the brewing process.

Figure 1.6. Representative Fuqu Baijiu products from China.

The typical production process of Fuqu involves purification of strains, culturing in small triangular flasks and culturing in large triangular flasks to obtain seed fuqu for further cultivation, which is inoculated with the ingredients to obtain Fuqu. The Fuqu cultivation generally can be achieved using the so-called qu-plate, curtain or ventilation method. Fuqu can be used in almost all flavor types of Baijiu production. The typical characteristics of the Fuqu Baijiu-making process are short fermentation time, high utilization rate of grain and over 70% fresh Baijiu yield. Fuqu technology in making Baijiu was widely promoted in 1955 through the 'Yantai Baijiu Brewing Operation Method'. However, due to the insufficient flavor of pure bran Baijiu, the brewing process of Daqu combined with Fuqu is generally adopted in the production process to ensure the fullness and perfection of the Baijiu body style, such as Sesame flavored Baijiu, in which Daqu and Fuqu are combined and used as saccharifying and fermenting starters.

With the continuous improvements of Baijiu knowledge and technological means, the pure strains inoculated with Fuqu also expanded from molds to bacteria and aroma-producing microorganisms. Fuqu is especially suitable for the production of high-quality Baijiu in the cold areas of northern China.

Fuqu is widely used in the brewing processes of Jingzhi Baijiu, Bandaojing Baijiu, Meilanchun Baijiu and Caoyuanwang Baijiu.

7. Mixed-qu Baijiu

Mixed-qu Baijiu as shown in Figure 1.7 is brewed with more than a single Qu as the saccharifying and fermenting starter. The mixed-qu may be achieved in two different ways. The mixed-qu can be a combination of 'Daqu & Xiaoqu' or 'Daqu & Fuqu', the traditional mixed-qu. It can also be a Qu directly inoculated with multiple microorganisms with different functions during brewing.

The mixed-qu Baijiu made with 'Daqu & Xiaoqu' uses sorghum, corn, wheat and/or other grains as ingredients. Daqu and Xiaoqu are used together. Daqu is used for generating flavor substances, while Xiaoqu works more effectively for saccharification and fermentation power. This combination takes advantage of the two kinds of Qu. The Dong flavor type of Baijiu is produced using this technology.

The technology of Baijiu-making using Daqu & Fuqu has been applied in the production of Light, Sesame and Sauce flavor types of Baijiu. The Light flavor type Baijiu is fermented by adding Fuqu into the grains after the fermentation of Daqu. The Sesame flavor type of

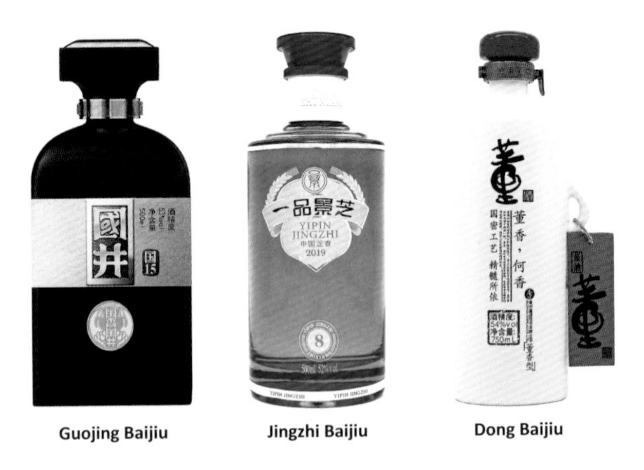

Figure 1.7. Representative Fuqu Baijiu products from China.

Baijiu is produced using a combined Fuqu (90%) and Daqu (10%). The Sauce flavor type of Baijiu is fermented using sequential fermentation of Daqu followed by that of Fuqu. The combination of 'Daqu & Fuqu' improves the mouthfeel, and the overall taste and aroma of Baijiu as compared to that made using Fuqu alone.

The multifunctional Qu consists of multiple strains of mold, yeast and bacteria selected due to their strong activities of protease and/or amylase and/or aroma production. These multiple strains may be cultivated together to obtain the desired Qu, followed by enri with possible utilization of up-to-date biotechnologies. The obtained multifunctional Qu is expanded and eventually used in Baijiu brewing. Modern biotechnology may be utilized when necessary.

Dong Baijiu, Yanghe Mianrou Baijiu, Jingzhi Baijiu and Bandaojing Baijiu are all made using a mixed-Qu.

8. Solid-state Fermented Baijiu

According to the national standard GB/T 15109-2008 of the People's Republic of China, solid-state fermented Baijiu is defined as a product with inherent style characteristics: it is made from grain(s) as the ingredients through solid (or semi-solid) saccharification and fermentation, followed by distillation, aging and blending, without edible alcohol and/or flavor substances from non-alcoholic fermentation.

The solid-state fermented Baijiu is mainly produced through solid-state or semi-solid-state fermentation. Among the six distilled alcoholic beverages in the world, Chinese Baijiu is the only category which is produced by the solid-state fermentation process.

The solid-state fermentation process uses sorghum, rice, glutinous rice, maize, wheat and other cereals as the starting materials. The cereals are cooked with water, cooled and mixed with a Jiuqu, and fermented in a pit. Daqu, Xiaoqu or Fuqu converts the grain starch to sugar (saccharification), which is then fermented into alcohol in the pits. After fermentation, the solid or semi-solid mixture containing alcohol, named Jiupei, is distilled in a Zengtong to obtain the base Baijiu. After aging and careful blending, a commercial solid-state fermented Baijiu is eventually produced from the base Baijiu.

Solid-state fermentation technology has been utilized in the production of the Chinese traditional flavor types of Baijiu, including the Strong, Sauce, Light, Mixed, Feng, Laobaigan, Sesame, Herblike (Dong), Te and Fuyu Baijiu products. Figure 1.8 shows several representative products made by solid-state fermentation technology.

Semi-solid fermentation can generally be carried out in two ways: fermentation after saccharification of the cultured microorganisms and simultaneous saccharification and fermentation.

The technology of saccharification followed by fermentation is a typical production process of the Rice flavor type Baijiu. Rice is used as the ingredient, and is soaked in water, steamed and pasted, ventilated and cooled to moderate temperature, followed by mixing with Xiaoqu for solid-state saccharification. After 18–24 hours, water is added for liquid-state fermentation and semi-liquid fermentation for a total of 5–7 days. Sanhua Baijiu from Guilin of Guangxi Province and Xiangshan Baijiu from Quanzhou City are typical representatives of the products from this technique.

Figure 1.8. Representative Baijiu products made by solid-state fermentation technology.

The process of simultaneous saccharification and fermentation is a typical procedure of the Chi flavor type of Baijiu and belongs to the traditional liquid-state fermentation technology. The Chi flavor type of Baijiu is an aroma type derived from the Rice flavor type. In the production of the Chi flavor type of Baijiu, steamed rice is inoculated with Xiaoqu, and mixed with water in a jar for saccharification and fermented at the same time. Yubingshao Baijiu from Guangdong Province is a typical example of the products from this technique.

Because of the unique open production technology and the diversity of microorganisms involved in Baijiu-making, the solid-state fermented Baijiu is rich in trace components, which are the key factors determining the aroma, taste and style of Baijiu.

9. Liquid-state Fermented Baijiu

In March, 2018, the national standard *Terminology and Classification of Alcoholic Beverage (Draft for Comments)* was released by the National Technical Committee for Alcohol-making Standardization (NTCAS). The standard defined the Baijiu produced by a liquid-state fermentation method as a refined spirit that is made from grains after liquid saccharification, fermentation, distillation, with or without the possible addition of edible alcohol fermented from grains and the non addition of coloring, flavoring and/or aroma components from a separated fermentation procedure.

The difference between Baijiu by liquid-state and solid-state methods is primarily the production process, and the quality of the two Baijiu products should not be compared. Baijiu from a liquid-state process is also a pure grain-derived Baijiu. The Chi flavor type Baijiu as shown in Figure 1.9, one of the 12 typical Baijiu in China, is made by a liquid-state fermentation procedure.

Liquid fermentation is an advanced processing technique, and the Baijiu by liquid fermentation is becoming the mainstream process. Compared with the traditional solid fermentation, liquid fermentation has more advantages in mechanization, automation, intelligence,

Figure 1.9. Yubingshao Baijiu products made by liquid-state fermentation technology.

efficiency and overall cost. Brandy, Vodka, Whiskey and some other famous distilled alcoholic drinks in the world are all made by liquid fermentation methods.

10. Raw Baijiu

Raw Baijiu (RB), also known as base Baijiu as shown in Figure 1.10, refers to the fresh Baijiu obtained by a fermentation and distillation procedure and without any post-processing treatments such as blending.

RB products should not be consumed directly due to the high alcohol concentration generally ranging from 55 to 75 degrees, as well as the uncoordinated flavor of the newly distilled RB, which has a strong sense of pungency and dryness. An aging process is therefore required. The aging process can not only get rid of the low boiling point substances in RB while the ethanol molecules and water molecules are mixing together, thus reducing the activity of ethanol molecules, but also makes the Baijiu taste more mellow and soft by cultivating the subtle chemical interactions among alcohol, aldehyde, acid and other components to generate new flavor compounds.

RB is also classified into three parts: the head, the middle and the tail. The head RB is the alcohol-water mixture collected at the beginning of distillation and generally has a relative greater alcoholicity; the

Figure 1.10. Raw Baijiu from a Baijiu-making enterprise.

tail RB has a relative lower alcoholicity and is collected at the end of distillation; and the middle RB is collected in the middle of the distillation and has the best quality.

The commercial 'Raw Baijiu' is generally blended Baijiu product, but not a real 'RB' by definition. However, compared with the real RB, the commercial 'Raw Baijiu' has a more harmonious body and palatability.

11. The Blending of Baijiu

The commercial Baijiu is made by blending RBs from different production batches and years, followed by diluting with water to get the alcohol content according to the brand and quality standards.

According to the *Terminology of Chinese Spirits Industry* (GB/T 15109-2008), 'blending', a technical term in the Baijiu industry, is defined as mixing base Baijiu specimens with different aromas, tastes and styles in certain proportions in order to ensure the consistent quality and sensory characteristics of all bottled commercial Baijiu for the brand.

The quality of Baijiu from different fermentation pits may be different due to the possible differences of raw materials, environment and time regardless of the fermentation technology. Even in the same fermentation pit, the quality of Baijiu may vary depending on the production seasons, fermentation time and operation personnel. Without blending, the quality of the bottled Baijiu may differ from one lot to another. It is the blending that ensures the consistent quality and palatability of Baijiu from the same brand and series.

To ensure a consistent standard and quality of Baijiu, retain the traditional style and maintain the color, aroma, taste and quality characteristics, blending is a necessary step for Baijiu manufacturers.

'Adding edible alcohol, food flavors and water during Blending' is prohibited for Baijiu. 'Blending' is limited to mixing Baijiu of different production batches, years and flavors according to certain requirements. The crude spirits used in the blending are all made by traditional fermentation methods.

Ethanol and water account for about 98% of Baijiu, while minor components account for the other 2%. It is these minor components that determine the style and quality of Baijiu. The minor components in Baijiu from different production batches and years are different. Blending helps the minor components in the proper proportion meet the quality standards and the style characteristics.

Blending is indispensable in the process of Baijiu manufacturing. All commercial Baijiu has gone through blending. It is well accepted that 'aroma-producing depends on fermentation; aroma-improving comes from distillation; Baijiu-forming comes from blending'. Blending technique is a crucial step for the final quality and sensory characteristics in the Baijiu-making process.

Traditional blending is done by the masters according to their experiences and creative tests. Figure 1.11 depicts the traditional blending scene in a Baijiu-making enterprise in China. With the advancement of Baijiu chemistry research and the rapid development of artificial intelligence technology, possible blending by computers may greatly improve blending efficiency and Baijiu quality.

Figure 1.11. Traditional blending process in a Baijiu-making enterprise in China.

12. Cellar

The cellar (pit) is one of the fermentation containers used in the process of solid-state Baijiu brewing, where the saccharification and fermentation of the grain is hosted. In general, a cellar is a pit dug in the ground, which is filled with grain for saccharification and fermentation (shown in Figure 1.12) before moving the grain for distillation to obtain Baijiu.

During the construction of this pit, one had to pay attention to the local congenital, natural conditions such as the topography and soil quality. The materials used in a pit vary from place to place in China. Generally speaking, the pits for making the Strong flavor type Baijiu are mud pits, and the bottom and surrounding sides of the pits are covered with yellow mud. The pits for producing the Sauce flavor type Baijiu are constructed with stones, and the surrounding sides are built with stones and the bottom is covered with yellow mud. The pits for making the Light flavor type Baijiu can be constructed using cement and bricks.

Figure 1.12. The cellar used in the process of Yingjiagong Baijiu, a kind of solid-state fermented Baijiu.

The Strong flavor type Baijiu brewing masters often say that 'A millennium old cellar for a million years of Jiuzao (distillers' grains), and the aging of a cellar the great quality of its Baijiu'. Every distillery regards its own old cellar as important, some of which may have lasted several hundred years. The pit mud in the cellar is rich in long-term domesticated diverse species of Baijiu-making microorganisms. These microorganisms transform starch, protein and other substances in grain into alcohol and different varieties of flavor substances. The longer the cellar is used, the more microbial colonies are accumulated and the more stability there is in the cellar environment, which results in the most favored brewing process of Baijiu containing more trace components and a consistent quality.

It is not difficult to note that cellars are very important and mysterious because of their essential microorganisms besides being the place for grain fermentation. If the secrets of microorganisms in the cellars can be revealed through modern technology, any container can be an excellent cellar for high-quality Baijiu.

13. Underground Jar

The underground jar (commonly called Digang, as shown in Figure 1.13) is also a kind of Baijiu fermentation container. As its name implies, it is a ceramic vat buried in the ground. It is mainly used in the production of the Light and Laobaigan flavor types of Baijiu. From the production history of Baijiu, the use of a vat has a longer history than pits, and it is mainly used in the areas with obvious continental climate. The typical representatives of the Light and Laobaigan flavor types of Baijiu are Fen and Hengshui Laobaigan Baijiu, respectively, and both are fermented in Digang with a common depth of 1.2 meters and a diameter of 0.8 meters.

Changes in the temperature and humidity during fermentation of the raw materials in the Baijiu-making process may have an important impact on microbial proliferation and metabolism, thus affecting the production of aroma substances. It was the wisdom of the ancient workforce to adopt the method of placing the jars underground for fermentation according to local conditions. Digang can keep the Jiupei (fermented grains) warm, while the proper air permeability of the jars is conducive to the entry of trace oxygen in the soil into the jar to help facultative bacteria to proliferate rapidly. As fermentation progresses, the temperature of

Figure 1.13. The Digang (underground jar) used in the brewing process of the Light flavor type Baijiu.

fermented grains increases gradually. At this time, Digang can transmit heat to the surrounding soil, so as not to inhibit the activity of microorganisms due to high-temperature stress. The temperature drops in the late period of fermentation, and Digang and the surrounding soil preserve the heat to keep the temperature of the fermented grains from dropping too quickly and reducing the microbial activity. It is well accepted that Digang is very helpful to maintain the temperature requirement of 'slow beginning, stiff middle and slow ending' in the fermentation process. In addition, compared with the mud pit, the ceramic vat separates the fermented grains from the soil and does not allow the soil bacteria to alter the fermented grains, which is more conducive for the pure characteristic flavor and clean aftertaste of the Light flavor type of Baijiu.

14. Peach Blossom Earthen Jar

The earthen jar, called 'Weng' in Chinese, is a pottery container as shown in Figure 1.14, which has been used to store grain and water since ancient times. It is also a traditional container for brewing Baijiu. In *Shuowen Jiezi*, a graphics book explaining the definition and derivation of Chinese characters, the earliest meaning of '酉(You, means alcohols)' is alcohols because it looks like the 'Weng' used for brewing Baijiu. Dongpo Su's '*Honey Alcohols Song*' described his Baijiu-making process as 'gentle boiling as fish bubbling on Day 1, ebullient ting and shinning the dance honey on Day 2, and the aroma reaching far when opening the earthen jar on Day 3'. The alcohol-making container used by Dongpo Sun is Weng.

Jingzhi Town, one of the three ancient towns in Shandong Province, is also a famous Baijiu-making town. It has a tradition of 'harvest in autumn, store in winter and open cellar for brewing in spring'. A unique coarse sand earthen jar is used as the Baijiu-making container. The earthen jar has certain permeability, which is conducive for exchanging gas and heat between the Jiupei inside and the external environment. In ancient times, Jingzhi Town prevailed in Baijiu-making, had many distilleries and made Baijiu according to the

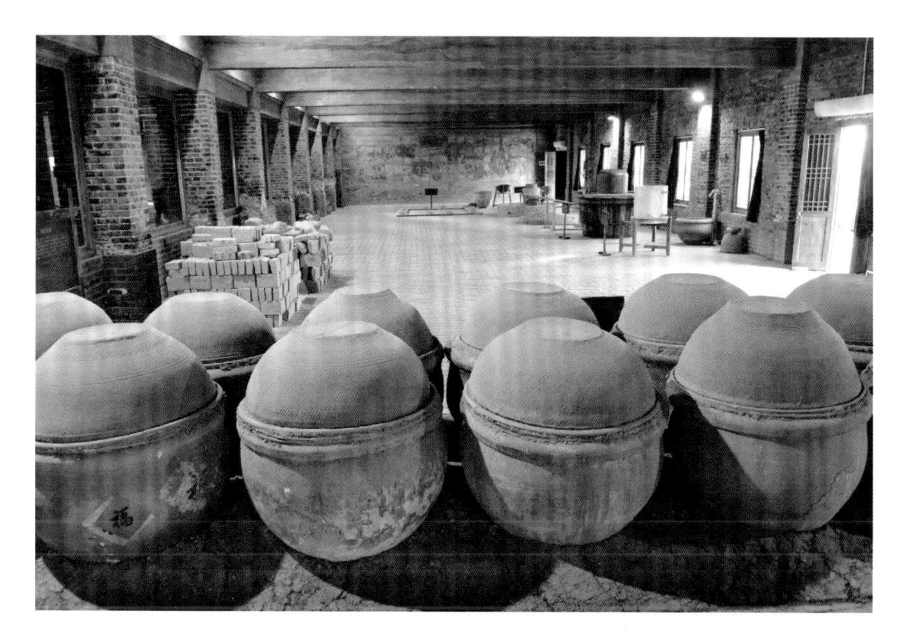

Figure 1.14. The earthen jar, Weng, used for brewing Baijiu in Jingzhi Town, Shandong Province.

seasons and environmental conditions. The Baijiu makers select high-quality grains to prepare Jiupei in the autumn harvest season. In winter, the Jiupei is kept in soil ponds for fermentation, which is conducive to moisturizing and heat preservation. In spring, the Jiupei is moved into the pottery jar as everything in nature recovers and microbial reproduction is vigorous, so that the temperature of Jiupei can meet the fermentation law of 'slow beginning, middle stiff, slow down at the end', which is conducive to improving the quality of Baijiu. When the peach blossoms are in full bloom, the Jiupei is removed from the pottery jar and steamed to obtain Baijiu. At this time, the Baijiu is rich in flavor, and pleasant to drink. It is historically called the 'peach blossom earthen jar' Baijiu.

Jingzhi Town still retains the distillery site of South Campus, which was one of the '72 pots' of Jingzhi at the time. It shows the ancient techniques of stone grinding, manual trampling and Tianguo distillation. The pottery jars of peach blossom earthen jar Baijiu in

Jingzhi Town in ancient times are also completely preserved, so that people can appreciate the style of the peach blossom earthen jar in the Baijiu brewing process during the Ming and Qing Dynasties.

15. Zengtong

Zengtong, as shown in Figure 1.15, is a steamer barrel, or a solid-state distillation device used in Baijiu production for traditional solid-state fermentation in China. Apparently, 'aroma generation depends on fermentation, and flavor improvement depends on distillation'. Distillation is an important step to extract ethanol and other aroma components from Jiupei to obtain Baijiu. The importance of Zengtong is self-evident.

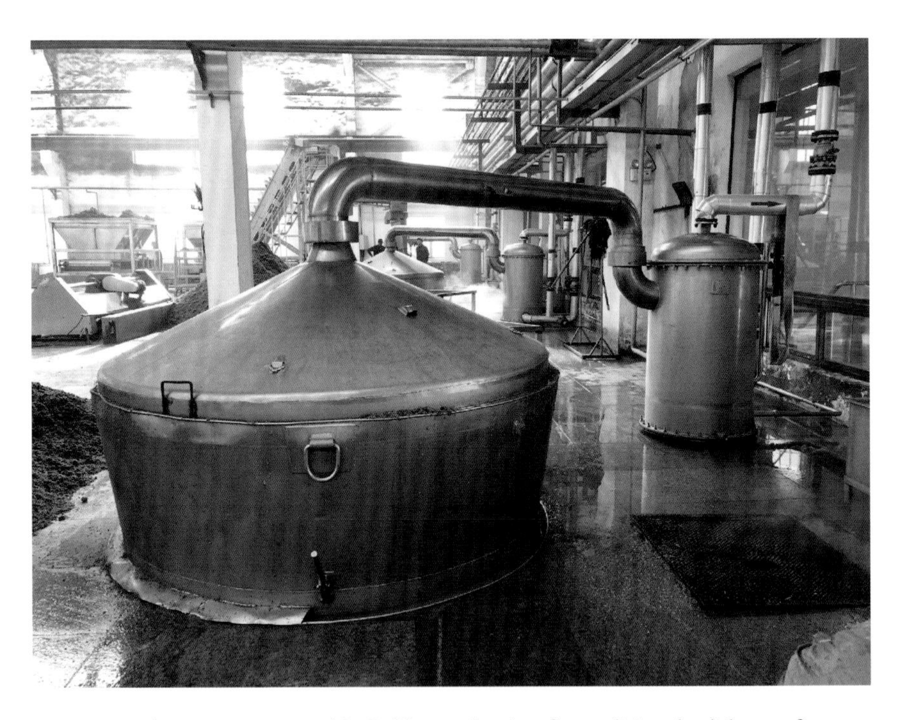

Figure 1.15. Zengtong used in Baijiu production for traditional solid-state fermentation technique.

Compared with other distillation devices in the world, Zengtong is designed for distillation of solid grains. The design is particularly special. The traditional Zengtong is made of wood, but now stainless steel is mostly used. Zengtong is composed of a barrel body, a steamer cover and a bottom pot. The bottom pot is filled with the Huangshui (the yellow serofluid) and the tail of Baijiu, to provide steam for steaming. The top of the bottom pot is a cone-shaped barrel body. The diameter of the upper and lower barrel body is about 2 and 1.8 meters, respectively, and the height is about 1 meter. It looks like a huge flowerpot. A sieve plate separates the bottom pot and the barrel body. When steaming, the Jiupei is evenly filled on the sieve plate in the barrel. Under the action of the steaming in the bottom pot, the temperature of the Jiupei in the steamer keeps rising. The volatile components in different layers of Jiupei are concentrated and extracted through continuous gasification and condensation. Jiupei with only 4% alcohol content can be distilled and concentrated to form the base Baijiu with about 60% alcohol content. At the same time, many trace volatile components and very few non-volatile components produced by microbial fermentation can also be extracted into the distilled fresh Baijiu. In the process of distillation using a Zengtong, the filling technology, the loose degree of Jiupei, the amount of steam and so on all have great impacts on the quality of Baijiu. It can be said that steaming Baijiu is a technical job requiring much experience.

Zengtong is not only used to steam Baijiu but is also used for steaming the grains. According to the requirements of different Baijiu production processes, some grains are steamed separately, whereas some are steamed together with Jiupei. The purpose of steaming the grain is to gelatinize the starch and kill germs in the original grain materials to carry out the fermentation at a later period. However, Zengtong, as a distiller of traditional solid-state Baijiu brewing, has a low distillation efficiency. Therefore, further research and development are required for more efficient distillation equipment and techniques on the basis of inheriting traditional technology for modernization of Baijiu production.

16. Jiuhai

In ancient times, the large-capacity alcohol-drinking vessel was named 'Jiuhai', which looked like the sea full of liquor. *Chinese Dictionary* explains that the drinking vessel is called 'Jiuhai' because of its large size. In fact, besides referring to drinking vessels in general, Jiuhai is also a unique container for alcohol storage that has nearly a thousand-year history.

Jiuhai, as show in Figure 1.16, looks like a large basket made of thorns and rattan on the outside, or a wooden box, and the inside is pasted with 'blood material' and paper or cotton cloth layer by layer. The 'blood material' is a plastic protein salt made of animal blood (usually pig blood) and lime, which forms a semi-permeable film when exposed to alcohol. This film allows gas permeation and exchange, but the liquid cannot pass through. With the help of the characteristics of the semi-permeable membrane,

Figure 1.16. Jiuhai used in the Tianyoude Baijiu-making process.

Jiuhai is very magical as it can ensure the process of gas exchange between the Baijiu body and the outside environment without leakage. Gas exchange can affect the rate of redox, molecular association and esterification reactions in the Baijiu, and plays a unique role in Baijiu aging. A small amount of dissolution of the inner coating material in Jiuhai during storage also may alter the style of Baijiu.

Jiuhai is the crystallization of the wisdom of the working people of ancient China. The capacity of Jiuhai ranges from hundreds of kilograms to tens of tons. Jiuhai is the largest traditional container for alcohol storage in the global brewing industry. The most famous existing Jiuhai is the Xifeng Jiuhai in Shanxi Province. In 2017, 12 ancient Jiuhai vessels in Xifeng distillery were recognized as a cultural relic.

The Jiuhai in Xifeng distillery looks like a large basket made of thorns. Its diameter is usually 2–2.5 meters, and it looks like a huge jar. The Baijiu makers in Xifeng distillery collected the vitex from Qinling Mountains in autumn, and wove it into a large basket before the water had disappeared. The inner wall of the Jiuhai was covered with a white cotton cloth with blood and lime as the binder, after which it was pasted with hemp paper. Hundreds of layers of hemp paper were pasted, and each layer was dried naturally before the next paper layer could be pasted. Finally, the surface was coated with a formula containing egg white, rapeseed oil and beeswax in a certain proportion for a smooth and well-sealed container. The pasted Jiuhai should be dried naturally until February 2nd of the following year before it can be used. A larger Jiuhai may take several years to make.

Besides Xifeng distillery, Jiuhai are used for Baijiu storage in many distilleries, such as Shanyan Baijiu in Liaoning Province, Taibai Baijiu in Shanxi Province, Jinhui Baijiu in Gansu Province, Daquanyuan Baijiu in Jilin Province and a few others.

Jiuhai is also used in Qinghai Huzhu Barley Wine Co., Ltd to store the highland barley Baijiu.

17. Jar

The history of the jar, a container widely used to store alcoholic beverages, is as long as the Chinese alcohol. The original alcohol jar was made of pottery, but porcelain jars are widely used now. Compared with the stainless steel tank, the porcelain jar occupies more space and costs more, but it is better in gas permeability and heat preservation. The other microelements in the porcelain jar may accelerate the aging of Baijiu. Famous distilleries prefer porcelain jars (shown in Figure 1.17) as the containers for base spirits. The newly distilled raw Baijiu is usually stored in porcelain jars for years of natural aging before blending and bottling.

The distilleries of both Baijiu and Huangjiu use big jars to store crude spirits, and general consumers may use different kinds of smaller jars at home. Full of unique cultural details, the alcohol jars contribute important components to the Chinese alcohol culture and tradition. Alcohol jars with calligraphy, drawings, allusions and porcelain arts may reflect the spirit culture, auspicious culture and regional culture of that time. The alcohols may be given auspicious meanings through the auspicious designs painted on the jars. For example, calabash patterns are usually painted on the alcohol jars or the alcohol jars are made in a calabash shape to express good wishes because the pronunciation of calabash is almost the same as that for 'happiness and richness' in Chinese, and the large number of seeds in calabashes implies 'more offspring, more blessings'.

Figure 1.17. Porcelain jars used as the containers for base Baijiu.

18. Drinking Vessels

The drinking vessel is the container(s) for serving spirits. In modern times, small-sized distributors and handleless cups made of glass or porcelain are usually used to serve Baijiu, which are shown in Figure 1.18 on the left (A) and right (B) upper corners, respectively. With the development of the alcohol-making industry and society, the drinking vessels vary greatly. The pottery vessels in the Neolithic Age, the bronze vessels in the Shang and Zhou Periods, the lacquer vessels in the Han Dynasty and the porcelain vessels in the Sui and Tang Dynasties were all drinking vessels for alcohol. According to the raw materials, drinking vessels are classified into vessels made of natural materials (such as animal horns or calabashes), pottery, bronze, porcelain, lacquer, gold and silver, jade, crystal, glass and plastics. According to their functions, drinking vessels are classified into the vessels to store, to warm and to drink alcohol in general.

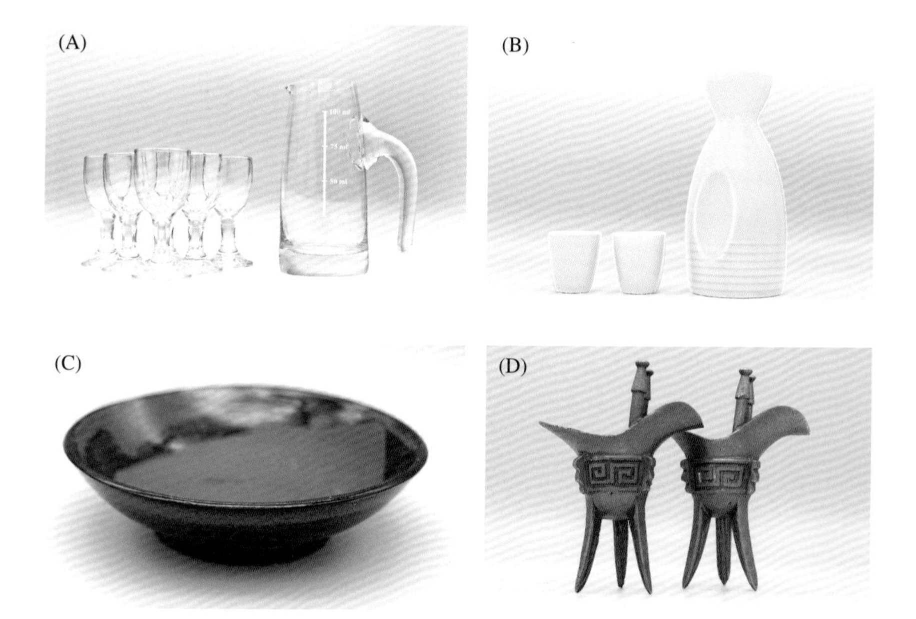

Figure 1.18. Different kinds of drinking vessels: (A) and (B) the handleless cups made of glass and porcelain, respectively; (C) the bowl as drinking vessel; (D) the delicate drinking vessels with special designs made of bronze.

In ancient times, a vessel with multiple purposes was common. The original drinking vessels were the same as those for serving foods such as bowls, as shown in the left lower corner (C) in Figure 1.18, or other vessels with large openings. Dedicated alcohol-drinking vessels made of pottery appeared in the Longshan culture of the Neolithic Age. Drinking vessels made of bronze were unprecedented, prevailing from the Shang and Zhou Dynasties to the Spring and Autumn Period and the Warring States Period due to the improvement of the bronze-making techniques. It is still shocking for us to see those delicate ancient drinking vessels with special designs, as shown in the right lower corner (D) in Figure 1.18 (a copy), exquisite Taotie (a mysterious beast representing glutton) patterns and high artistic value in museums. The vessels to store alcohol include '尊 (Zun)', '壶 (Hu)', '鉴 (Jian)', '斛 (Hu)', '觥 (Gong)', '瓮 (Weng)', '瓿 (Bu)', '斝 (Jia)', '盉 (He)' and '彝 (Yi)'. Jia and He are also used to warm alcohol. Jue, Gong, Zhi and Piao are used to drink alcohol. Human beings select different vessels and follow different ritual codes for different occasions, seasons and guests. For instance, *The Book of Rites* describes that 'in a sacrifice rite, the noble uses Zhi whereas the humble uses Jiao when serving alcohols'.

Not as fragile as pottery or as complex as a bronze one, drinking vessels made of lacquered wood were popular in the Qin and Han Periods. However, they were only used to drink alcohol not to warm or store alcohol. The primitive drinking vessels made of porcelain originated in the Shang and Zhou Periods, and rapidly developed in the Han Dynasty because of their low cost and durability. Porcelain drinking vessels dominated and led the market for a long time. The development and manufacture of porcelain drinking vessels reached a peak in the Ming and Qing Periods with a large number and high quality. Due to the availability of raw materials, drinking vessels made of jade or gold were exclusive to the upper class from the ancient times to the late Qing Dynasty.

Nowadays, drinking vessels made of glass, of high quality and inexpensive, have become the market mainstream. Bottles are the most common vessels to store alcohol. Glass bottles and porcelain bottles are used to store Baijiu. Every Baijiu brand develops its own designed bottles considering its style and culture. Drinking vessels are classified into cups, Zhong (handleless cup), Zhan, bowl, etc., according to the local customs of drinking. The development and advancement of drinking vessels are associated with the economic and technological development of human society and may promote the development of the alcohol culture. Every drinking vessel is the collective wisdom of the working people of its times.

Chapter 2
History

19. The History of Chinese Alcoholic Drinks

Chinese alcoholic drinks have a long history and China is one of the earliest countries in the world to prepare alcoholic drinks. Chinese alcohol-making techniques date back to the Jiahu cultural period, 9000 years ago, and the raw materials were rice, honey, grapes, hawthorn berries and so on. The earliest alcoholic drinks in China would be rice, fruit and honey liquors. The history of Chinese alcoholic drinks is much longer than that mentioned in legends when 'alcohols were made by Di Yi' or 'alcohols were made by Kang Du'.

The history of beer in the world is over 8000 years old. The oldest document about alcoholic drinks was about the brewing techniques of beer for sacrifice engraved in clay plates by Babylonians in about 6000 B.C. The beer brewing techniques in China might date back 5000 years, which was concluded from researching and analyzing the residue in pottery such as the alcohol-drinking vessel excavated from the Mijiaya site in Xi'an City.

The history of Chinese Huangjiu has lasted about 7000 years, which has been proved by the large amount of unhusked rice and pottery similar to alcohol-drinking vessels excavated from the Hemudu site in Hemudu Town, Yuyao City, Zhejiang Province.

The history of grape wine in the world is at least 7000 years old. Musallas from Xinjiang Province with a history of over 3000 years is the oldest grape wine in China. The 'fine grape wine' from the lines 'Fine grape wine in luminous cups of jade: to drink I want but the summoning Pi-pa on horseback is played' in *Song of Liangzhou* by Han Wang might be the description of Musallas.

The history of Baijiu in China is at least 2000 years old which has been proved by the vessels used to distill Baijiu excavated from the tomb of Haihun Marquis of the Han Dynasty.

The oldest real alcohol found in the world was the 3000-year-old grape wine excavated in Samari, Iran. The oldest real alcohol found in China was the royal alcohol made from grains excavated in Xi'an City.

Rice alcohol (Mijiu), Huangjiu and Baijiu are all unique kinds of alcoholic drinks in China. The order of their invention is rice alcohol first, then Huangjiu, and Baijiu last. The time intervals among them are long. More than 9000 years ago, rice alcohol was invented; after 2000 years of development, Huangjiu was obtained by filtering rice alcohol; and after another 5000 years, Baijiu with greater alcohol content was invented after distillation of Huangjiu accompanied by the invention of distillation devices.

The Chinese character '酒 (Jiu, means alcohols)' and other characters connected with '酒 (Jiu)', such as '醴 (Li)', 尊 (Zun)' and '酉 (you, means alcohols)', in the inscriptions on bones or tortoise shells are the evidence of the long history of alcohol. The records of alcohol in the literature and history are too numerous to mention. The following lines in the *Book of Songs*, the first collection of poems in China, are also evidence of the long history of Chinese alcohol culture: 'peeling jujube in August; harvesting rice in October; making alcohols from them; best wishes for a long life' and 'good alcohols makes one drunk; tasty food makes one full.'

20. The History of Huangjiu

Stemming from and prospering in China, Huangjiu is one of the most ancient traditional fermented alcohols in the world, and one of the three ancient alcohols in the world along with grape wine and beer.

As to the origin of Huangjiu, there are many stories. One story is that Di Yi or Kang Du first made Huangjiu. The story that it originated in the Huangdi Period is also well known. A more mysterious proposal is that 'there are alcohol stars in the universe, they first prepared alcohols for the world and heaven'. The well-accepted proposal by Tong Jiang, a master and government officer in the West Jin Dynasty, is that Huangjiu was created by chance. Tong Jiang, in his Chapter *Alcohol Orders*, suggested a story that leftover rice dumped near the mulberry trees might have mixed with cooked wheat and millet and fermented naturally to form the first alcohol.

The residue of grain is essential for the generation of Huangjiu. The large amount of cultivated rice and the pottery similar to alcohol vessels excavated from the Hemudu site (shown in Figure 2.1) in the eastern Zhejiang Province are evidence that the history of fermented alcohol from grains in China might be at least 7000 years old. Huangjiu is officially recorded in *the Book of Songs*. There are more than 50 records about rice growing and alcohol brewing and drinking, which date back 2800 years. There were also many

Figure 2.1. The pottery 'He' in a bird shape, dated 7000 years old, excavated from the Hemudu site. ('He' is an ancient Chinese drinking vessel used to mix alcohol and water.)

written records about Huangjiu during the Period of Goujian, the king of Yue. It was recorded in *The History of Yue* that 'giving birthto a boy is rewarded with two pots of Huangjiu and a dog whcreas that for a girl is rewarded with two pots of Huangjiu and a pig'. This was the policy encouraging birth issued by the king of Yue in order to increase the number of soldiers and laborers for the country. The records in *Lv's Commentaries of History* also mentioned alcohol to express the king's determination to gain support from the people: 'if the delicious food was not enough to distribute to the common, the king would not eat it; the king would pour spirits in the river to share it with the common.' The records in The *Spring and Autumn of Wu and Yue* that 'alcohols were served when the king was leaving' prove that Huangjiu might have been used in official state events. The record in *The History of Han Dynasty* that 'two buckets of unpolished rice and one bucket of Qu generate six buckets of alcohols' is the earliest record in the world about the proportion of the raw materials and products in alcohol-making.

There are many varieties of Huangjiu from different locations. The popular producing areas include Zhejiang, Jiangsu, Fujian, Hubei, Shanxi and Shandong Provinces and Shanghai City in China.

21. The History of Baijiu

Baijiu, also called 'Shaojiu', is a distilled alcoholic beverage unique to China. Distilled alcoholic beverages, such as Baijiu, Brandy, Whiskey and Vodka, are spirits made from the distillation of fermented products and contain a greater alcohol content after distillation.

The history of distilled alcohol is closely connected to the invention and use of vessels for distilling alcohol. The earliest distilled alcohols were made by ancient people from Ireland and Scotland, who used pottery vessels to distill the alcohols, leading to the initial preparation of Whiskey.

The earliest time when Chinese Baijiu was created has not been determined. The well-accepted original times include the Yuan Dynasty, Song Dynasty, Tang Dynasty or Eastern Han Dynasty. In fact, the origin of Chinese Baijiu can possibly date back to the Western Han Dynasty, more than 2000 years ago, earlier than the abovementioned eras.

No doubt, Baijiu was initiated after the invention and use of vessels to distill alcohol. Figure 2.2 shows a distillation device used in the Jin Dynasty.

Haihun Marquis was a title of nobility in the Western Han Dynasty with a four-generation inheritance (from 68 B.C. to 8 A.C.). A bronze vessel for distillation consisted of a bronze still and a barrel, the same as the distillation vessel used in traditional distillation. One such was excavated from the alcohol storehouse in the tomb of He Liu (from 98 B.C. to 59 B.C.), the first Haihun Marquis of the newly construct District, Nanchang City, Jiangxi Province. As it was excavated from an alcohol storehouse in the tomb, it is probable that it was used for alcohol distillation.

In the period of Haihun Marquis, Whiskey, the earliest distilled liquor from the Occident, had been made, while it had been 7000 years since the production of Huangjiu in China. Therefore, it is easy to understand that the vessels founded in the tomb of Haihun Marquis were used to distill alcohol.

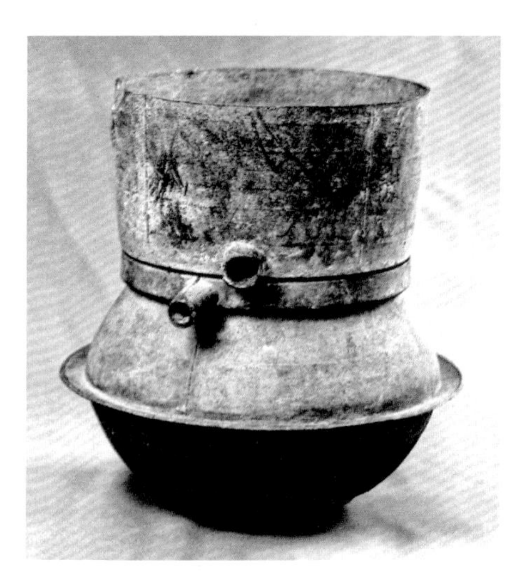

Figure 2.2. A distillation device used in the Jin Dynasty.

22. The Conflict between the Chu and Han, and the Banquet at Hongmen

The banquet at Hongmen is a well-known historic dinner where treachery, to murder an invited guest, was planned in the Hongmen village near Xianyang, the capital city of the Qin Dynasty, in B.C. 206, during the period of peasant uprisings and the conflict between the Chu and Han countries. The host was Yu Xiang, the chief general of the Chu, while the primary guest was Bang Liu, the lord of Pei County and the chief general of the Han. Zeng Fan, the under father of Yu Xiang, suggested and planned the murder of Bang Liu at the banquet. Bang Liu followed the advice of his counselor Liang Zhang, who realized the possible risk of the planned murder, and apologized to Yu Xiang voluntarily to show his kindness and weakness. During the banquet, Zeng Fan dropped several hints about assassinating Bang Liu, but Yu Xiang hesitated because of his reputation. Without other choices, Zeng Fan asked Zhuang Xiang to perform a sword dance to entertain everyone at the banquet and seize a possible opportunity to assassinate Bang Liu in his seat. At that very moment, Bo Xiang, Yu Xiang's uncle, drew his sword to dance with Zhuang Xiang, and protect Bang Liu with his body, stopping Zhuang Xiang from carrying out the assassination, and therefore, saving Bang Liu. Bang Liu left quietly and quickly with his bodyguard Kuai Fan without saying goodbye, and his gifts and apology were presented to Yu Xiang by Liang Zhang, who stayed a little longer. Figure 2.3 depicts the scene of the Banquet at Hongmen.

After the banquet at Hongmen, Bang Liu launched the four-year-long war between the Chu and Han countries during which the power of Bang Liu grew from weak to strong and turned the tide to defeat Yu Xiang. After the victory, Bang Liu established the Han Dynasty and was its first emperor. Those outstanding figures in the later times always regarded Yu Xiang as a negative example of a failure due to not getting the job done completely. Zedong Mao wrote in his Qilv poem (an eight-line poem with seven characters in each line) entitled *People's Liberation Army Has Occupied Nanjing* that 'with

Figure 2.3. The hand-drawn story happened at the banquet at Hongmen in China (The hand drawing is courtesy of Song Zhang, BTBU).

power and to spare we must pursue the tottering foe; not ape Yu Xiang, the conqueror loving the idle fame'.

'Banquet at Hongmen' usually refers to a banquet or an event with no good intentions nowadays. Some Chinese idioms are also derivatives of 'Banquet at Hongmen' such as 'not commit the slightest offence against people', 'toilsome service but distinctive merits', 'to draw up a few rules', 'being foods on a chopping board', 'the sword brandished by Xiang Zhuang aims the murder of the lord of Pei County' and 'all sides of Chu music'.

23. Kuangyin Zhao's Removal of Military Power by Serving Alcohol

The story of removal of military power by means of cups of alcohol happened in the early Song Dynasty. Kuangyin Zhao, the general of the Later Zhou Dynasty, led a military mutiny with Pu Zhao in Chenqiaoyi in A.D. 960. Kuangyin Zhao was crowned emperor by other generals who put the imperial yellow robe on him. It is known as 'the military mutiny in Chenqiao' or 'draped with the imperial yellow robe'. Then, the troops of Kuangyin Zhao marched on the capital, Kaifeng. The senior generals guarding the capital including Shouxin Shi and Shenqi Wang opened the gate to let Zhao's troops in and forced the emperor of the Later Zhou Dynasty to abdicate the throne. Kuangyin Zhao established the Song Dynasty by changing the title of the dynasty into 'Song'.

At the very beginning of the Song Dynasty, Kuangyin Zhao learnt a lesson from the demise of the Later Zhou Dynasty, tightened his grip on the imperial troops and took some measures to strengthen the central authority. At first, Kuangyin Zhao did not pay too much attention to his generals such as Shouxin Shi, but Pu Zhao, the Prime Minister, constantly reminded him of the risk of another 'the military mutiny in Chenqiao' or usurpation by other generals, and got his attention. After considering and evaluating the facts that the emperor's power had been weak while that of the subordinates was strong since the late Tang Dynasty, and there had been twelve emperors with eight different last names during the decades of the Five Dynasties and Ten Kingdoms, Kuangyin Zhao decided to conduct a legendary historical play of removing the military powers of the generals.

On the ninth day of July of the lunar year, the second year of the Jian Long period (the year of 961), Kuangyin Zhao invited Shouxin Shi, Huaide Gao and several other senior generals for a banquet after a meeting. During the feast, Shouxin Shi, Huaide Gao and the other generals were threatened and bribed, which led to their resignations, submitted voluntarily the following day to give up military power due to personal health conditions. Kuangyin Zhao accepted this with pleasure and also fulfilled his promises, offering them positions out of the capital and great wealth. This story is called 'Removal of Military Powers by Serving Alcohols', because it was done peacefully during a feast while drinking alcohol. Figure 2.4 depicts the scene of Kuangyin Zhao's removal of military power by serving alcohol at a banquet.

Now, the idiom 'Removal from Military Powers by Serving Alcohols' refers to removing the military powers of generals without difficulty.

Figure 2.4. The hand-drawn story of Kuangyin Zhao's removal of military power by serving alcohol at a banquet (The hand drawing is courtesy of Song Zhang, BTBU).

24. Crossing the Chishui River Four Times and Moutai Baijiu

Chishui River, 523 km in length, also called Dashe River, Anle River and Chihui River in the ancient times, is an upper tributary of the Yangtze River. It originates from Zhenxiong County of Yunnan Province, flows through Chishui City, Guizhou Province and Sichuan Province, and combines with the Yangtze River at Hejiang County of Luzhou City, Sichuan Province.

Chishui River is known as a river of good spirits because the Chishui River valley has been a production base of good liquors from ancient times. Moutai and Lang, the two famous Sauce flavor type Baijiu brands in China, are both made along Chishui River. Zhen Zheng once wrote in his *Moutai Village* that 'Moutai produces the best Baijiu in Guizhou; salt is transported into the Chishui River'. Guohua Zhang from the Qing Dynasty, also stated that 'Moutai Baijiu is so good that people from Yunnan, Sichuan, Guizhou and

Hunan Provinces come to enjoy it; Moutai was sold in the market thousands of miles away and people all appreciated it'.

'Crossing the Chishui River Four Times' was a significant act of mobile warfare by the central red army during the long march after the Zunyi Meeting from January 15–17, 1935. 'Crossing the Chishui River Four Times' was the masterpiece of Zedong Mao and played an important role in the Chinese revolution (Figure 2.5). 'Crossing the Chishui River Four Times' further established and consolidated the leadership of Zedong Mao in the party and red army, turned the unfavorable tide at the beginning of the long march and laid the foundation for victory of the long march.

The 'Crossing the Chishui River Four Times' campaign lasted three months. The border area of Sichuan, Yunnan and Guizhou Provinces where the red army deployed was the golden triangle region of Chinese Baijiu consisting of Luzhou, Sichuan Province, Yibin, Sichuan Province, and Zunyi, Guizhou Province, where Moutai Baijiu, Xi Baijiu, Luzhou Laojiao Baijiu, Lang Baijiu, Wuliangye Baijiu, Jiannanchun Baijiu, Tuopai Baijiu, Shuijingfang Baijiu and other famous brands of Baijiu are made. Erlangtan Town, Gulin County, Luzhou, the home of Lang Baijiu, is the place where the second and the fourth crossings of 'Crossing the Chishui River' took place. Moutai Town, the home of Moutai Baijiu, is the place

Figure 2.5. The oil painting of the 'Crossing the Chishui River Four Times' campaign.

where the third crossing of 'Crossing the Chishui River' happened.

After the foundation of the People's Republic of China, Enlai Zhou named Moutai Baijiu the alcohol for national ceremonies and events because the central red army stopped by Moutai Town during the fourth 'Crossing the Chishui River'. Enlai Zhou once told Richard Nixon at a banquet in 1972 that 'Moutai Baijiu made a great contribution to the victory of long march'. When the red army arrived at Moutai Town during the long march, they used Moutai Baijiu to clean up wounds, reduce inflammation and keep out the cold. Those who loved drinking enjoyed the good Baijiu and the others filled Moutai in bottles to rub on their feet and invigorate the circulation of blood during the long march. The following lines written by Yi Chen at a banquet with Yanpei Huang in Nanjing in 1952 also proved it: 'what we drink when meeting again in Nanjing is what we had used to wash feet during the long march.'

One of the authors of this book described 'Crossing the Chishui River Four Times', Moutai Baijiu and Lang Baijiu in his *Feelings Expressed in Yan'an* during his time at the China Executive Leadership Academy in Yan'an in October, 2015, stating that 'At the Zunyi Meeting, Zedong Mao became a member of the Executive Committee. Crossing the Chishui River Four Times was a nice story for Chinese spirits. The aroma of Moutai Baijiu lingered alongside the Chishui River. Lang Baijiu was popular in Erlang Town, Luzhou. It is De Zhu who named Luzhou as 'the City of Baijiu'. The famous Baijiu were of high quality and large quantity. Baijiu was so great that the officers and men regarded it as treasure. Baijiu was used to rub feet when tired during the long march.'

25. The Eight Famous Alcohol Brands

At the First National Alcohol Appraisal Conference (NAAC) in 1952, eight brands of famous Chinese alcohols were awarded the prize of recognition of 'famous'. These included four Baijiu brands, one Huangjiu brand, two grape wine brands, and one fruit wine brand.

Alcohol has been the mainstay of the national economy and state revenue since the ancient times. When the People's Republic of China was established in 1949, the national economy was in tatters and there were almost no modern industries. To revitalize the alcohol industry, the First NAAC was held in Beijing in 1952 at the initiative of Prime Minister Enlai Zhou and China General Administration of Monopoly. 103 alcohol brands were displayed at the conference including 19 Baijiu brands, 16 grape wine brands, 9 brandy brands, 28 blended alcohol brands, 24 other alcohol brands and 7 medicinal alcohol brands. The brands awarded had to meet four requirements:

1. The brand should be of high quality and meet the standard of superior alcohol and the hygienic standard.
2. The brand should earn favorable comments and be popular with most people.
3. The brand should have a long history and should still be marketed nationally.
4. The brand should have a unique manufacturing technique, local characteristics, and should be hard to imitate.

According to the analytical data from the Beijing Testing Factory (Beijing Distillery, Beijing Red Star Co., Ltd. nowadays), the requirements mentioned above and recommendations from experts, eight famous alcohol brands were awarded and listed as given below in Table 2.1 and Figure 2.6.

Table 2.1. Information on the eight famous alcohol brands.

Category	Brand name	Business name
Baijiu	Moutai Baijiu	Moutai distillery, Guizhou
	Luzhou Laojiao Tequ Baijiu	Luzhou distillery, Sichuan
	Fen Baijiu	Xinghuacun distillery, Fenyang, Shanxi
	Xifeng Baijiu	Xifeng distillery, Shaanxi
Huangjiu	Shaoxing Jiafan Huangjiu	Shaoxing winery, Zhejiang
Grape wine	Red Rose Grape Wine	Zhangyu winery, Yantai, Shandong
	Vermouth Wine	Zhangyu winery, Yantai, Shandong
Fruit wine	Special Fine Brandy	Zhangyu winery, Yantai, Shandong

Figure 2.6. The eight famous alcohol brands in the First NAAC.

The four Baijiu brands were the earliest 'Four Famous Brands' in the history of Chinese Baijiu. The First NAAC not only laid a great foundation for the alcohol appraisal but also awarded the eight famous brands, which played a significant role in promoting alcohol production and quality improvement in China.

26. The Eight Famous Baijiu Brands

At the Second NAAC, eight famous Baijiu brands were awarded a prize, including Wuliangye Baijiu, Gujinggong Baijiu, Luzhou Laojiao Tequ Baijiu, Quanxing Daqu Baijiu, Moutai Baijiu, Xifeng Baijiu, Fen Baijiu and Dong Baijiu.

The Second NAAC was held by the Food Industry Bureau of the Ministry of Light Industry in Beijing in October, 1963. Each province, autonomous region and municipality carefully evaluated and selected their alcohol products representing those on the commercial market as required. The samples with profiles were co-sealed with a signature and co-submitted by the provincial, autonomous regional or municipal bureau of light industry and its bureau of commerce. After several runs of local selection, 196 alcohol brands in 5 categories, Baijiu, Huangjiu, grape wine, beer and fruit wine, were recommended for the final competition by 115 companies from 27 provinces, autonomous regions or municipalities.

The appraisal was led by the appraisal committee and all the judges followed the rules strictly. Baijiu, Huangjiu, fruit wine and beer were tested and judged separately. Liqueur with Baijiu as the base spirit was tested and judged by the Baijiu committee, while liqueur with ethanol as the base spirit was evaluated by the fruit wine committee.

Every judge rated the brands on a scale of 0–100 according to the overall color, aroma and taste independently and made comments for their sensory evaluation. Numbered with codes, the samples were selected through group elimination, preliminary assessment, reassessment and final assessment. The winners were selected according to the scores.

Among 75 Baijiu samples, eight Baijiu brands were named Famous Baijiu and nine Baijiu brands were named Quality Baijiu according to the 15 judges. The information on these eight famous Baijiu brands is listed in Table 2.2, and the pictures are shown in Figure 2.7.

During the Second NAAC, a professional assessment committee for Baijiu was established, professional judges were recommended, coding was used, assessment rules were established and the management measures for famous and quality Baijiu brands were also discussed. It was great progress scientifically compared with the First NAAC.

Table 2.2. Information on the eight famous Baijiu brands.

Number	Brand name	Business name
1	Wuliangye Baijiu	Wuliangye distillery, Yibin, Sichuan
2	Gujinggong Baijiu	Gujinggong distillery, Bo County, Anhui
3	Luzhou Laojiao Tequ Baijiu	Luzhou distillery, Luzhou, Sichuan
4	Quanxing Daqu Baijiu	Chengdu distillery, Sichuan
5	Moutai Baijiu	Moutai distillery, Guizhou
6	Xifeng Baijiu	Xifeng distillery, Shaanxi
7	Fen Baijiu	Fenjiu distillery, Xinghuacun, Shanxi
8	Dong Baijiu	Dongjiu distillery, Zunyi, Guizhou

Figure 2.7. The eight famous Baijiu brands in the Second NAAC.

27. The Famous and Quality Alcohol Brands in China

The NAAC has been held five times since the People's Republic of China was established in 1949. The appraisal of the quality alcohol brands began since the Second NAAC.

The Second NAAC was held by the Ministry of Light Industry in Beijing in 1963. Among 75 Baijiu samples, eight brands were awarded the title of the State Famous Baijiu and nine brands were awarded the title of the State Quality Baijiu.

The Third NAAC was held by the Ministry of Light Industry in Dalian, Liaoning Province, in 1979. Among 106 Baijiu samples, eight brands were awarded the title of the State Famous Baijiu and 18 brands were awarded the title of the State Quality Baijiu. It was the first time that Baijiu brands were evaluated and compared within the groups assigned by their flavor type, alcohol content, raw materials and sacchariferous agents.

The Fourth NAAC was held by the Food Industry Association of China (FIAC) in Taiyuan, Shanxi Province, in 1984. Among 148 Baijiu samples, 13 brands were awarded the title of the State Famous Baijiu and 27 brands were awarded the title of the State Quality Baijiu. The Evaluation Committee of the National Quality Award (ECNQA) awarded these 13 State Famous Baijiu brands the Golden Prize of National Quality Food and the 27 State Quality Baijiu brands the Silver Prize of National Quality Food.

The Fifth NAAC was held by FIAC in Hefei, Anhui Province, in 1989. Among 361 Baijiu samples, 17 brands were awarded the Golden Prize (State Famous Baijiu) and 53 brands were awarded the Silver Prize (State Quality Baijiu). During this conference, the Famous and Quality Baijiu brands of the previous NAAC were reevaluated and the submitted brand samples were also evaluated.

The information on the State Famous Baijiu brands in the First–Fifth NAAC is summarized in Table 2.3 and the Baijiu products are shown in Figure 2.8.

Table 2.3. Information on the State Famous Baijiu brands in the First–Fifth NAAC.

Number	Brand name	Flavor style	Manufacturer	NAAC No.
1	Moutai Baijiu	Sauce flavor	Moutai distillery, Guizhou	1,2,3,4,5
2	Fen Baijiu	Light flavor	Xinghuacun Fenjiu company	1,2,3,4,5
3	Luzhou Laojiao Tequ Baijiu	Strong flavor	Luzhou distillery	1,2,3,4,5
4	Xifeng Baijiu	Other (Feng flavor now)	Xifeng distillery	1,2,4,5
5	Wuliangye Baijiu	Strong flavor	Wuliangye distillery	2,3,4,5
6	Gujinggong Baijiu	Strong flavor	Gujing distillery, Bozhou	2,3,4,5
7	Quanxing Daqu Baijiu	Strong flavor	Quanxing distillery, Chengdu	2,4,5
8	Dong Baijiu	Other (Dong flavor now)	Dongjiu distillery, Zunyi	2,3,4,5
9	Jiannanchun Baijiu	Strong flavor	Jiannanchun distillery, Mianzhu	3,4,5
10	Yanghe Daqu Baijiu	Strong flavor	Yanghe distillery	3,4,5
11	Shuanggou Daqu Baijiu	Strong flavor	Shuanggou distillery	4,5
12	Huanghelou Baijiu	Strong flavor	Wuhan distillery	4,5

Table 2.3. (*Continued*)

Number	Brand name	Flavor style	Manufacturer	NAAC No.
13	Lang Baijiu	Sauce flavor	Langjiu distillery, Gulin	4,5
14	Wuling Baijiu	Sauce flavor	Wulin distillery, Changde	5
15	Baofeng Baijiu	Light flavor	Baofeng distillery	5
16	Songhe Baijiu	Strong flavor	Songhe distillery, Luyi	5
17	Tuopai Baijiu	Strong flavor	Tuopai distillery, Shehong	5

Figure 2.8. The State Famous Baijiu brands in the First–Fifth NAAC.

At the Fifth NAAC, the Silver Prize (State Quality Baijiu) was awarded to the following: Harbin Teniang Longbin Baijiu (Daqu starter, Sauce flavor), Sichuan Xufu Daqu Baijiu (Daqu starter, Strong flavor), Hunan Deshan Daqu Baijiu (Daqu starter, Strong flavor),

Hunan Liuyanghe Xiaoqu Baijiu (Xiaoqu starter, Rice flavor), Guangxi Xiangshan Baijiu (Xiaoqu starter, Rice flavor), Guangxi Sanhua Baijiu (Xiaoqu starter, Rice flavor), Jiangsu Shuanggou Teye Baijiu (Daqu starter, Strong flavor, low alcohol content), Jiangsu Yange Daqu Baijiu (Daqu starter, Strong flavor, low alcohol content), Tianjin Jin Baijiu (Daqu starter, Strong flavor, low alcohol content), Henan Zhanggong Daqu Baijiu (Daqu starter, Strong flavor), Hebei Yingchun Baijiu (Fuqu starter, Sauce flavor), Liaoning Lingchuan Baijiu (Fuqu starter, Sauce flavor), Liaoning Dalian Laojiao Baijiu (Fuqu starter, Sauce flavor), Shanxi Liuquxiang Baijiu (Fuqu starter, Light flavor), Liaoning Lingta Baijiu (Fuqu starter, Light flavor), Harbin Laobaigan Baijiu (Fuqu starter, Light flavor), Jilin Longquanchun Baijiu (Fuqu starter, Strong flavor), Inner Mongolia Chifeng Chenqu Baijiu (Fuqu starter, Strong flavor), Hebei Yanchaoming Baijiu (Fuqu starter, Strong flavor), Liaoning Dalian Jinzhouqu Baijiu (Fuqu starter, Strong flavor), Hubei Baiyunbian Baijiu (Daqu starter, Mixed flavor), Guangdong Shiwan Yubingshao Baijiu (Xiaoqu starter, Chi flavor), Shandong Fangzi Baijiu (Fuqu starter, other flavor), Hubei Xiling Tequ Baijiu (Daqu starter, Mixed flavor), Heilongjiang China Yuquan Baijiu (Daqu starter, Mixed flavor), Sichuan Ere Daqu Baijiu (Daqu starter, Strong flavor), Anhui Kouzi Baijiu (Daqu starter, Strong flavor), Sichuan Sansu Tequ Baijiu (Daqu starter, Strong flavor), Guizhou Xi Baijiu (Daqu starter, Sauce flavor), Sichuan Sanxi Daqu Baijiu (Daqu starter, Strong flavor), Shanxi Taibai Baijiu (Daqu starter, other flavor), Shandong Kongfujia Baijiu (Daqu starter, Strong flavor), Jiangsu Shuangyang Tequ Baijiu (Daqu starter, Strong flavor), Heilongjiang Beifeng Baijiu (Fuqu starter, other flavor), Hebei Congtai Baijiu (Daqu starter, Strong flavor), Hunan Baishaye Baijiu (Daqu starter, other flavor), Inner Mongolia Ningcheng Laojiao Baijiu (Fuqu starter, Strong flavor), Jiangxi Site Baijiu (top grade, Daqu starter, other flavor), Sichuan Xiantan Daqu Baijiu (Daqu starter, Strong flavor), Jiangsu Tanggou Tequ Baijiu (Daqu starter, Strong flavor), Guizhou An Baijiu (Daqu starter, Strong flavor), Dukang Baijiu (Daqu starter, Strong flavor), Sichuan Shixiantaibai Chenqu Baijiu (Daqu starter, Strong flavor), Henan Linhe Tequ Baijiu (Daqu starter, Strong flavor), Sichuan

Baolian Daqu Baijiu (Daqu starter, Strong flavor), Guizhou Zhen Baijiu (Daqu starter, Sauce flavor), Shanxi Jinyang Baijiu (Daqu starter, Light flavor), Jiangsu Gaogou Tequ Baijiu (Daqu starter, Strong flavor), Guizhou Zhuchun Baijiu (Fuqu starter, Sauce flavor), Guizhou Meijiao Baijiu (Daqu starter, Strong flavor), Jilin Dehui Daqu Baijiu (Fuqu starter, Strong flavor), Guizhou Qianchun Baijiu (Fuqu starter, Sauce flavor) and Anhui Suixi Teye Baijiu (Daqu starter, Strong flavor). The first 25 Baijiu samples were the State Quality Baijiu brands from the previous NAAC that had been reassessed in this NAAC. The other 28 Baijiu samples were newly awarded the title of State Quality Baijiu brands.

Chapter 3
Culture

28. The Cultural Connotations of Alcohol

Culture is the combination of the material and spiritual wealth generated in the development of human society and history. Alcohol is a cultural component. The cultural connotations of Chinese alcohols are abundant, and are greater than any other alcohols in the world.

At the material level, alcoholic drinks have a long history. The history of Chinese alcohols is over 9000 years. The history of Huangjiu is over 7000 years, and the history of Baijiu is over 2000 years. There are many different styles and flavor types of alcohols. As the saying goes, 'the best Huangjiu is from Shaoxing in the south and Dai County in the north'. Other significant Huangjiu producers are located in Fujian, Shandong, Shanghai, etc. These Huangjiu products have some similarities, but differ from each other due to the different ingredients, manufacturing techniques, their aroma and taste. The 12 flavor types of Baijiu come from distilleries all over China and every Baijiu brand has its own style and characteristic. Many famous Baijiu products also have their own unique culture.

Alcohols also contribute much and in many ways at the spiritual level. People in ancient times believed that alcohol was able to connect humans with deities, so it was sacrificed to the heaven and earth, ghosts and gods for good fortune from the very beginning. The primary utilization of alcohol was to worship gods and ancestors, and mourn the dead. The Chinese character '奠 (Dian)' is derived from an image of an alcohol pot placed on a table. 'Dian' is defined as to show respect and to mourn the dead by offering sacrifice.

If there is no alcohol, there is no banquet. Alcohol is indispensable in almost all social events including state banquets, family feasts, festival feasts, celebration parties, birthday parties, parties in honor of teachers, wedding feasts, farewell parties, receptions for friends and guests, and events like seeing friends off, as depicted in the drawing in Figure 3.1.

Alcohols have been closely associated with many famous writers since ancient times as depicted in Figure 3.2. Literary giants such as Bai Li, Fu Du, Juyi Bai, Qingzhao Li, Mu Du and Xiu Ouyang in the Chinese history all enjoyed alcohol with enthusiasm and left many poems and well-known stories about alcohols.

The alcohol culture is rich and varies with different expressions among the communities, both elegant and vulgar. We should promote and develop the outstanding alcohol culture components and discard the parts that are not good.

Figure 3.1. The hand-drawn story of farewell of friends. (The hand drawing is courtesy of Song Zhang, BTBU.)

Figure 3.2. The hand-drawn story of enjoying alcohols with friends. (The hand drawing is courtesy of Song Zhang, BTBU.)

29. The Etiquette about Drinking Alcohol

China is known as 'the State of Etiquette' and everything goes according to the rules. Drinking alcohol also has its own rules for good manners. The alcohol etiquette enhances the sense of ritual and makes drinking an official, solemn and civilized social event.

With its large population, China has a long history and diverse cultures. Except the non-alcohol-drinking populations, the 56 ethnic groups of China may have their own drinking etiquettes that can be very different and cannot be described in detail. According to the saying 'an etiquette is to be humble to yourself and respectful to others' in *the Book of Rites*, it is necessary to follow the local drinking codes of the hosts, which is in accordance with the phrase, 'when in Rome, do as the Romans do'.

Invitation for a drink has its rules. The invitation for a meal and/or alcohol drinking shall be early enough, so that your guests can better plan and prepare for it. An invitation three days in advance is an 'invitation to someone for a gathering for a meal and drink', two days in advance is more to 'ask someone for a meal or drink', and on the same day is considered 'dragging someone for a drink' according to the traditional Beijing rules.

There are also rules to be kept in mind when presenting alcohol as gifts. Alcohol can be a gift for a marriage engagement, which implies an everlasting and unchanging relationship in Hebei Province. Four or eight bottles of alcohol are usually chosen, which means everything is fine.

It is important to figure out the seating order in advance because it may vary depending on the locations, the host, the guest, and the ages of the participating parties. Sitting at will is not a good idea. In general, the seat facing the door of a round table is for the chief host. The seat on the right side of the chief host is for the principal guest. Then, other hosts and guests can be seated at intervals as shown in Figure 3.3. In some other areas, the seat facing the door of a round table is for the principal guest and the host sits on the left side of the principal guest.

Figure 3.3. The hand drawing of the etiquette when drinking Baijiu and Huangjiu in China. (The hand drawing is courtesy of Zimei Zhao, BTBU.)

There are different etiquettes in toasting. Usually, the chief host proposes a toast to the guest first, followed by that from the guests. A short toast speech always goes with the toast. The general rules also depend on the locations and occasions. For instance, in many places in Shandong Province, the guests propose a toast after all the hosts have done it. The guests can also toast each other, and so can the hosts.

Clinking glasses is common in a toast. Respect is expressed by lowering your own glass below that of the other when clinking glasses in the toast as shown in Figure 3.3.

The custom of Henan Province that the host serves the alcohol for the guest but does not drink himself derives from a period of poverty in the past. At that time, the host would not have adequate amount of alcohol for all guests and hosts, and would make sure that the guests could have enough to enjoy.

As the saying goes, ‘one’s drinking manners reflect his virtue’, and the alcohol etiquette reflects one’s moral cultivation. The etiquette about alcohol is so that it aids better communication and friendship development, and helps avoid excessive intake or drunkenness. It is acceptable to replace alcohol by tea or water in an event involving alcoholic drinking. The view ‘other beverages could taste and feel alcohols as long as the friendship is real’ should also be recognized.

30. Alcohol and Filial Piety

Few people carefully think about the connections between alcohol and filial duty. In fact, our ancestors had given us the answer to it.

As the saying goes, ‘three treasures in a family are not as good as an elder’ or ‘the filial duty is the most important of all virtues’, and filial duty occupies an important position in the Chinese traditional culture. Carrying out filial duty is one of the traditional virtues in China. It is the wish of every family and descendant to bring the elders happiness and health. The 9th day of September in the lunar calendar, called ‘double nine’, is the Double Ninth Festival in Chinese traditional culture. This day is named the Day of Elderly People by the revised *Law on the Protection of the Rights and Interests of Elderly People* approved in a vote by the Standing Committee of the People's Congress of China on December 28th, 2012. Nine is the maximum single digit number and an auspicious number in the Chinese language. The combination of the number five and nine is 'Jiu Wu Zhi Zun' (the royal prerogative), which is only used by the emperor or the royal family.

Tian'anmen is the front gate of the Forbidden City in the Ming and Qing Dynasties. The main building of Tian'anmen is divided into the upper and lower parts. The upper part is nine rooms wide from east to west and five rooms deep from south to north. This is the combination of the numbers nine and five. There are five gates in the lower part, which also makes the formation of 'nine and five', symbolizing the dignity of the emperor. The middle gate is the biggest one, located on the central axis of the Forbidden City and only the emperor had the right to go through it in ancient times.

The pronunciation of the character 'nine' in Mandarin is the same as that of the characters 'permanence' and 'alcohols'. In addition, the pronunciation of 'nine and nine' is the same as 'long time' and 'alcohols and alcohols'.

'To make new alcohols on the 9th day of September in the lunar calendar' is a custom going back about 1000 years in Fang County, Hubei Province. Till now, Tucheng Village in Fang County, which is a well-known village far and near, still retains the custom of making Huangjiu in every household. 'To make new alcohols on the 9th day of September' is also the first line of the *Alcohols Divine Melody*, the episode of the movie *Red Sorghum* directed by Yimou Zhang. Today, some Baijiu enterprises often open or seal the Baijiu on this day because it not only shows the Chinese culture of good timing, geographical convenience and good human relations but also the close connections with elderly people.

The Chinese character '酵 (Jiao, means fermentation)' consists of two parts, the left and the right sides. The left part is the character '酉 (You, means alcohols)' which means alcohol in the inscriptions on bones or tortoise shells, and the right part is the character '孝' (Xiao, means filial duty) which clearly illustrates the relations between alcohol and filial duty in the Chinese character '酵 (Jiao)'.

What is the fermentation for? The left part of the character '酵 (Jiao)' is the answer, for alcohol-making.

What do we make alcohols for? The right part of the character '酵 (Jiao)' is the answer, for carrying out filial duty.

Why do we fulfill our filial duty by providing alcohol for the elderly people? According to traditional Chinese medicine, alcohol is same as a medicine and medicine comes from alcohol, and moderate drinking relaxes the muscles, stimulates the blood circulation and is good for health.

In Chinese history, Xian Li of the Tang Dynasty was the one who observed filial piety by serving alcohol and hid his capacities and bid his time. In 684, Xian Li was banished from the throne of Emperor by Zetian Wu, his mother, to become the Luling King in Fang County, Hubei Province. He not only liked Fang County Huangjiu but also paid tributes to Zetian Wu. In 699, Xian Li was reinstated as the crown prince and regained his power as the emperor in 705. Huangjiu in Fang County contributed a lot to the life of Emperor Xian Li.

Figure 3.4. The hand-drawn story of making alcohol for carrying out filial duty. (The hand drawing is courtesy of Zimei Zhao, BTBU.)

The character '酵 (Jiao)' also implies that minors should not drink alcohol. Although minors have not been prohibited expressly from drinking according to the existing laws in China, parents, schools and society still do not encourage minors to drink alcohol.

Moreover, the pronunciation of the character '酵' (Jiao) is not the same as that of the character '孝' (Xiao), but the same as that of the character '窖' (Jiao, means the alcohol pit).

31. Alcohol and Brewing

It is well known that Baijiu and Huangjiu are both made from grains. The original grains may include rice, glutinous rice, millet, sorghum and so on in ancient China. If there was no rice, there would be no grain. The left part of the Chinese character '粮 (Liang, means grains)' is the character '米 (Mi, means rice)', which shows the importance of rice as a grain.

The character '酿 (Niang, means brewing or alcohol-making)' clearly expresses the origin of alcohol and the relations between alcohol and rice and other grains. The Chinese character '酿 (Niang)' consists of two parts, the left part and the right side. The left part is the character '酉 (You)', the production of alcohol-making, which means alcohol. The right part is the character '良 (Liang, means goodness)' which shares the same pronunciation as the character '粮 (Liang)', the raw material used in alcohol-making.

In alcohol-making, the character '米 (Mi)' in the character '粮 (Liang)' was transformed into the character '酉(You)' which forms the character '酿 (Niang)'. With the transformation of the character '米 (Mi)', the '粮 (Liang)' is missing. The leftovers are the character '酉 (You)', and are mixed with water to form a new character '酒 (Jiu, means alcohols)'. Figure 3.5 shows the fermentation process (from grains to alcohol) in Digang used in the making of the Laobaigan flavor type Baijiu.

The sayings that alcohol is the conversion product of grains and alcohol is the derivative of grains are well grounded.

Although alcohol is the derivative essence of grains, Baijiu is not a simple aqueous solution of alcohol. More than 1874 minor components generated during the Baijiu fermentation progress have been

Figure 3.5. Digang used in brewing process of Laobaigan flavor type Baijiu.

found till 2017, which not only make Baijiu taste strong and mellow but also bring healthful functional factors.

32. Alcohol and Vinegar

Oriental vinegar originated from China. The documented history of vinegar-making in China is over 3000 years. The relation between vinegar and alcohol is clearly expressed by the character ‘醋 (Cu, means vinegar)’. The left part of the character ‘醋 (Cu)’ is the character ‘酉 (You, means alcohols)’, and the right part is the character ‘昔 (Xi, means the past)’. Vinegar stems from alcohol. The vinegar-making is carried out after the alcohol-making. Vinegar is the ‘past’ alcohol as the hand-drawn Figure 3.6 shows. Alcohol can be transformed into vinegar after 21 days. Today, Chinese rice vinegar, old vinegar and aged vinegar are still brewed in this way. The vinegars are made from Huangjiu and then fermented by acetic bacteria.

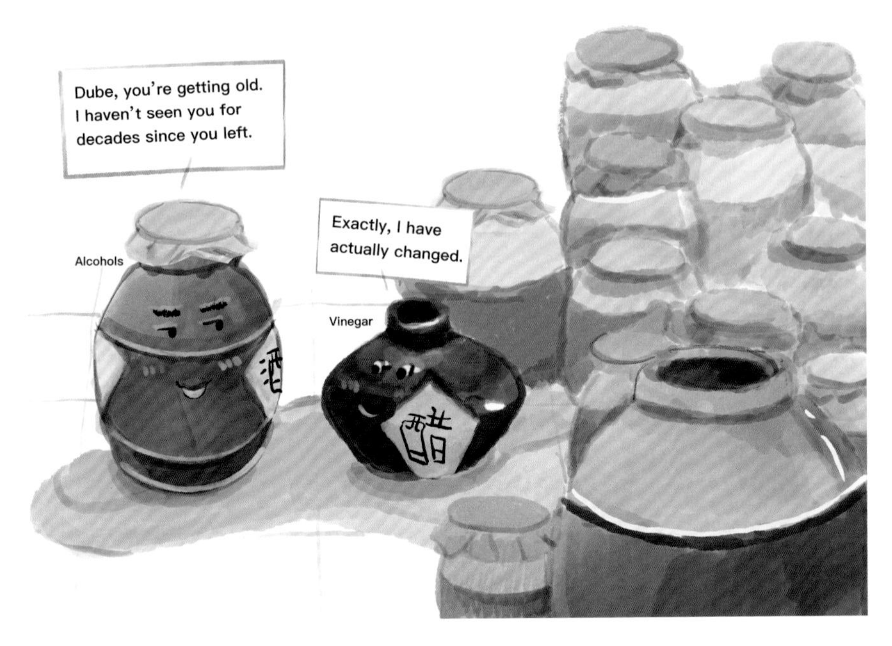

Figure 3.6. The hand-drawn story of vinegar, the 'past' alcohol. (The hand drawing is courtesy of Zimei Zhao, BTBU.)

It is said that 'aged alcohols and aged vinegars are the best'. The aged alcohols and aged vinegars not only taste better but are also healthier than freshly prepared alcohols and vinegars, because in the process of aging, elements such as alcohols, acids and aldehydes in alcohols and vinegars may change slowly and transform into components such as esters and acetals which contribute to better taste or improved effects to modulate human physiological functions.

Shanxi aged vinegar, Sichuan Baoning vinegar, Jiangsu Zhenjiang aromatic vinegar and Fujian Yongchun aged vinegar are the four famous vinegars in China. The affection for vinegar by Shanxi people is the same as that for chilies by Hunan residents. There is a saying in Shanxi that 'the men who do not like vinegar have no rich emotions; the women who do not like vinegar have no harmonious family; the children who do not like vinegar have no good performance at school; and the elderly people who do not like vinegar have no clear mind'.

A factory of aged vinegar in China, for example, in Gutian County, Fujian Province, adopts a more primitive and traditional way to make vinegar. Huangjiu can naturally become vinegar after being kept in ceramic jars for several years. The quality of vinegar that has been kept in the jars for more than ten years or several decades is better. Many jars from the Ming and Qing Dynasties are still used in this factory.

Interestingly, some countries in the world share the same way of vinegar-making as China by using alcohol as raw material. For instance, beers are used to make malt vinegar in Britain and the United States. Grape wines are used to make grape vinegar in France, Italy and Spain.

It is not unique, but has its counterpart. The word 'vinegar' in French also reflects the relationship between alcohol and vinegar. The prefix of the French word 'vinaigre' is 'vin' which means alcohol, red wine or grape wine in French.

33. Alcoholic Beverages, Alcohols, Aldehydes, Ketones, Acids and Esters

There is a component '酉 (You, means alcohols)' on the left of these Chinese characters '醇(Chun, means alcohols), 醛(Quan, means aldehydes), 酮 (Tong, means ketones), 酸 (Suan, means acids), and 酯 (Zhi, means esters)', indicating that they are all related to alcohol, which is true. By 2017, 1874 trace substances have been found in Baijiu, including 235 alcohols, 97 aldehydes, 140 ketones, 127 acids and 506 esters.

Alcohols, aldehydes, ketones, acids and esters are important substances that contribute to the flavor of alcoholic beverages. For example, ethanol, commonly known as alcohol, is necessary in all types of alcoholic beverages, and it is also a marker substance to distinguish soft drinks from alcoholic drinks. Beta-phenylethanol, with a sweet rose-like aroma, has been found in many types of alcoholic beverages; ethyl acetate, with a slightly fruity aroma, exists in all flavor types of Baijiu, and is also the key flavor substance of the Light flavor type Baijiu.

Many alcohols, aldehydes, ketones, acids and esters are important functional substances in alcoholic beverage. For example, caproic acid, heptanoic acid, octanoic acid, decanoic acid, lauric acid, myristic acid, stearic acid, oleic acid, ethyl linoleate and ethyl linolenate have the functions of inhibiting cholesterol biosynthesis.

Chemically, alcohols, aldehydes, ketones, acids and esters are also closely related to each other. Alcohols can be oxidized to aldehydes, ketones and acids. Alcohols react with aldehydes to form acetals, and alcohols react with acids to form esters. These changes may occur during the fermentation and aging process of Baijiu, and have an important impact on Baijiu flavor.

34. No Alcohol, No Banquet

In China, it is said that 'No alcohol, no banquet' and 'No fish, no banquet'.

The two sayings, long-standing and deeply ingrained, are the reflections of Chinese dietary culture, which can be easily answered by the Chinese character '宴 (Yan, means banquet)'. Another writing of the character '宴 (Yan)' is '醼 (Yan, means banquet)', which means to invite someone to attend a banquet for food, and to have a drinking party or banquet with someone.

The Chinese character '醼 (Yan)' consists of two parts, the left part and the right side. The left part is '酉 (You, means alcohols)', and the right part is '燕 (Yan, means fish)' referring to '燕鱼 (Yanyu, means fish)', a marine fish also called Spanish mackerel.

A probable explanation of 'No fish, no banquet' is that the pronunciation of the character '鱼 (Yu, means fish)' is the same as the character '余 (Yu, means surplus)' which means owning more than you need or abundant wealth in traditional Chinese culture. Therefore, mandarin fish is more popular in a banquet for its pronunciation in Chinese meaning wealth and surplus.

The Spring Festival is the Chinese New Year when all the family members get together and enjoy a reunion dinner called 'nian-ye-fan' as Figure 3.7 shows, which is usually a dinner on New Year's Eve, the last day of the lunar year, and is also the dinner in the early morning

Figure 3.7. The hand drawing of a hearty dinner on New Year's Eve in China. (The hand drawing is courtesy of Song Zhang, BTBU.)

of the first day of the Lunar New Year in some places in Jiaodong Peninsula. Fish has to be served at a nian-ye-fan in many regions, and usually is the last dish at the dinner. In some places, fish should not be eaten up or even not eaten at all to ensure 'owning more than you need'.

As for what kind of fish people eat, it depends on the region and what is available. No matter what kind of fish people eat, it represents 'surplus' which really matters.

35. *Book of Songs* and Alcohols

The *Book of Songs*, also known as *Poetry* or *Three Hundred of Poems*, is the earliest collection of poems in China, with 305 poems from the early Western Zhou Dynasty to the mid-Spring and Autumn Period (11th century B.C. to 6th century B.C.), which is shown in Figure 3.8. The *Book of Songs* is mainly composed of 'four sentence' poems, wherein a paragraph has four sentences containing two to

Figure 3.8. The cover of the *Book of Songs.*

eight words. The sentence 'By riverside a pair, of turtledoves are cooing. There is a maiden fair, whom a young man is wooing' was from the *Songs Collected South of the Capital, Modern Shaanxi and Henan,* Cooing and Wooing of the *Book of Songs.*

The *Book of Songs* is a classic book of Confucianism in China and one of the 'Five Classics' in the 'Four Books and Five Classics'. The poems in the *Book of Songs* were all lyrics that could be sung at that time. They were mainly used in ceremonies, entertainment and other occasions, where alcohol was naturally indispensable. The character '酒 (Jiu, mean alcohols)' appeared 63 times in the 305 poems in *Book of Songs*, which showed that alcohol was an indispensable product in both important ceremonies and banquets. It had become a custom in the high society at that time that no banquet might be hosted without alcohol.

The top-down enlightenment of the *Book of Songs* has always been valued. Confucius educated his disciples to read the *Book of Songs* as a criterion for their opinions and actions. Ethical and moral education is also reflected in the poems in the *Book of Songs*. The character '酒 (Jiu)' appeared 4 times and the character '醉 (Zui, means drunk)' appeared 13 times in *Revelry* of the *Book of Songs*, which depicted the guests thoroughly and profoundly from the orderly, courteous and dignified performance before drinking, the uncontrolled behavior and words after drinking, to the buffoonery and harmfulness after getting drunk. The following descriptions are brilliant: 'When guests begin to feast, they are gentle at least. When they're not drunk too much, they would observe the rite. When they have drunk too much, their deportment is light. They leave their seats and go capering to and fro. When they've not drunk too much, they are in a good mood. When they have drunk too much, they're indecent and rude. When they are deeply drunk, they know not where they're sunk. When they've drunk their cups dry, they shout out, brawl and cry. They put plates upside down. They dance like funny clown. When they have drunk wine strong, they know not right from wrong. With their cups on one side, they dance and slip and slide. If drunk they went away, the host would happily stay. But drunk they will not go, the host is full of woe. We may drink with delight if we observe the rite. Whenever people drink, in drunkenness some sink. Appoint an inspector and keep a register. But drunkards feel no shame, on others they'll lay blame'.

The ethical and moral education of the *Book of Songs* is also embodied in the following sentences: 'We've drunk wine strong and thank you grace. May you live long! Long live your race! We've drunk wine strong and eaten food. May you live long! Be wise and good!' *(The Book of Songs, Sacrificial Ode)*; 'Your wine is sweet and clear, and fragrant is your food. The spirits come to drink and eat, your blessing will be sweet' *(The Book of Songs, The Ancestor's Spirits)*.

The author of the *Book of Songs* is Jifu Yin, the Grand Tutor of the Zhou Dynasty. His hometown is Fang County, which is famous for making Huangjiu. At that time, Huangjiu was called 'Baimao', which is the earliest name of the Huangjiu produced in Fang County.

36. The Capital of Baijiu

There are many different styles and flavor types for Chinese Baijiu. The prosperity of the Chinese Baijiu industry has caused many cities to flourish such as Renhuai City in Guizhou Province, Fenyang City in Shanxi Province, Yibin City in Sichun Province and Suqian City in Jiangsu Province. Therefore, more and more cities where the Baijiu business is a major industry would like to name themselves the capital of Baijiu. In fact, the title 'Capital of Chinese Baijiu' is awarded by a group of experts from the China Light Industry Council and the China Alcoholic Drinks Association after systematically assessing the history and culture of Baijiu, the Baijiu-making environment, the scale of Baijiu enterprises and the local economic contribution rate of the Baijiu industry. To date, only two cities, Yibin, Sichuan Province, and Suqian, Jiangsu Province, have been awarded the 'Capital of Chinese Baijiu' officially in 2009 and 2012, respectively. The famous Baijiu brand Wuliangye is from Yibin, whereas Yanghe Daqu Baijiu and Shuanggou Daqu Baijiu are from Suqian. The title of 'Capital of Chinese Baijiu' is dynamic, and reevaluation is conducted every three years.

Located on the border between Sichuan Province and Yunnan Province, Yibin, as Figure 3.9 shows, belonging to Sichuan Province, is called the first city of great Yangtze River because the Yangtze River starts by merging Jinshajiang River and Minjiang River, and goes across northern Yibin. The origin of Baijiu-making in Yibin can date back to a period 4500 years ago, which has been proved by the pottery cups excavated from the Jiaohuayan site in Yibin. Baijiu-making among the common people originated in the Qin and Han Dynasties when Xunjiu and Jujiang liquors were popular. In the period of the Tang and Song, with the prosperity of tea-horse trade, the Baijiu-making skills and commodity transaction in Yibin developed rapidly.

Figure 3.9. The 'Capital of Chinese Baijiu', Yibin City. (Reproduced with the permission from Yichuan Gao).

The famous Baijiu brands included 'Chongbi Baijiu', 'Lizhilv Baijiu' and 'Yaozixuequ Baijiu'. The 'Lichuanyong', one of the large-scale Baijiu-making workshops, was established in the period of the Yuan and Ming. Zijun Deng from 'Lichuanyong' improved the technique of 'Wuliang' and designed the trademark of 'Wuliangye' in the period of the Republic of China. At the same time, other famous Baijiu brands as 'Jianzhuang Daqu', 'Tizhuang Daqu' and 'Tihu Daqu' appeared in Yibin. There are 66 Baijiu enterprises of considerable scale in Yibin. The famous enterprises are Wuliangye Group, Gaozhou Liquor, Xufu Liquor and Hongloumeng Liquor.

Located on the Yangtze River Delta, and the middle and lower reaches of Huai River and Yishusi Basin, Suqian, as shown in Figure 3.10, is under the jurisdiction of Jiangsu Province. The origin of Baijiu-making in Suqian can date back to the period of the Shang and Zhou, which has been proved by the alcohol vessels excavated from Shuanggou Town in Suqian. The brewing industry in Suqian began to flourish in the period of the Tang and Song. It was recorded in *Hongxian Records* that 'Confucianism was admired among the common, and drinking was popular in the period of Sui and Tang'. The lines during Songxizong's reign that 'the stars and the moon are all

Figure 3.10. The 'Capital of Chinese Baijiu', Suqian City.

asleep; while thousands of families are drinking' reflect the rapid development of the brewing industry at that time. Baijiu brands from Suqian were presented as tributes during the Qing Dynasty, and were praised as the best brand in the Jianghuai area because of the good taste. There were over 150 Baijiu enterprises in Suqian at the end of 2015. 'Yanghe Daqu' and 'Shuanggou Daqu' are two famous Baijiu brands in Suqian.

37. The City of Baijiu

Located in the southeast of the Sichuan Basin and at the junction of Sichuan, Yunnan, Guizhou and Chongqing Provinces, the hinterland of the Golden Triangle of Chinese Baijiu, Luzhou, the only City of Baijiu in China, is under the jurisdiction of Sichuan Province. The clear four seasons, the warm and moist climate, and the subtropical monsoon climate make Luzhou an appropriate place to make high-quality Baijiu. Figure 3.11 shows the Danwan Square of Luzhou.

Luzhou has a long history of Baijiu-making, starting from the period of the Qin and Han, booming in the period of the Tang and Song, and flourishing in the period of the Ming and Qing. The 'Baxiangqing Baijiu' from Luzhou, under the jurisdiction of Ba in the Zhou Dynasty, was presented as a tribute to the Zhou Dynasty. According to the antiques from Luzhou Museum, the pottery horn cups excavated in Luzhou were made in the period of the Qin and

Figure 3.11. Danwan Square of Luzhou City.

Han, 2000 years ago. The bronze vessel to warm alcohol with kylin patterns excavated from Naxi District, Luzhou, in 1986 was the auxiliary vessel for drinking in the Han Dynasty. Over 200 alcohol vessels excavated from Yinggoutou, Luzhou, in 1999 were used by the local residents in the period from the Sui and Tang to the Five Dynasties. Luzhou people in the Song Dynasty learned the ways to make alcohol and made 'Xiaojiu' and 'Dajiu', which is the embryonic form of Luzhou Baijiu. In the year of Taiding (1324) during the Yuan Dynasty, Huaiyu Guo from Luzhou was called the 'ancestor of Chinese Daqu alcohols' or 'father of Qu' because he created 'Ganchun Qu' to make alcohol. In the year of Wanli (1573) during the Ming Dynasty, Shu Chengzong, recognized as the 'ancestor of Strong flavor type Baijiu', established a 'Shujuyuan' Baijiu workshop, and developed Baijiu-making techniques for preparing Strong flavor type Baijiu such as 'the mud pit generates aroma' and 'the mixed Jiuzao with grains as the starting materials'. Since then, the production technology of Luzhou Daqu Baijiu has been gradually improving. In the Qing Dynasty, Baijiu-making workshops with a commercial shop in the front and a workshop in the back emerged in Luzhou. According to *The Records of Luzhou*, 'there were over 600 Baijiu-making workshops at the end of Qing Dynasty. The products were sold to Yongning and other places on the border of Guizhou Province. There were over 10 Daqu Baijiu-making workshops. The older the cellar, the more flavor the Baijiu is. Wenyongsheng and Tianchengsheng were famous for their aged cellars'.

The title 'City of Baijiu' was given by De Zhu. In 1916, De Zhu launched the war against Shikai Yuan with Er Cai from Yunnan to Sichuan, and their troops garrisoned in Luzhou. On New Year's Eve of that year, De Zhu wrote down a poem to express his feelings after drinking Luzhou Daqu Baijiu. He called Luzhou the City of Baijiu in this poem and this title was given to Luzhou then. One of the authors of this book wrote in his *Feelings Expressed in Yan'an* published on the official website of China Executive Leadership Academy in Yanan on October 22nd, 2015 that 'Luzhou is the City of Baijiu given by De Zhu. Its high quality and large quantity make the Baijiu famous'. The famous Baijiu brands in China 'Guojiao 1573', 'Luzhou Laojiao' and 'Lang' Baijiu are all made in Luzhou, the City of Baijiu.

38. The Hometown of Baijiu

China is the hometown of Baijiu. There are many hometowns of Baijiu in China, including Moutai Town, Renhuai, Guizhou Province; Xinghuacun Town, Fenyang, Shanxi Province; Tuopai Town, Suining,

Figure 3.12. Panorama of Moutai Town. (Reproduced with the permission from Yulong Luo).

Sichuan Province; Gujing Town, Qiaocheng District, Bozhou, Anhui Province; and Jingzhi Town, Anqiu, Shandong Province.

Located in the valley of Chishui River, Moutai Town is under the jurisdiction of Renhuai, Guizhou Province, as shown in Figure 3.12. The prosperity of Baijiu-making in Moutai Town was associated with the development of shipping in Chishui River in the middle and late Qing Dynasty. As the primary port in northern Guizhou at that time, Moutai Town was an important station through which the salt from Sichuan was delivered to Guizhou and the lead from Guizhou was delivered to Beijing, which stimulated the economic growth and the prosperity of the Baijiu-making industry. At present, Moutai Town is the leading producer of Chinese Sauce flavor Baijiu with over 300 Baijiu-making enterprises and over 4000 Baijiu trading companies among which Moutai Group is the most representative enterprise.

Located in the Zixia Mountain, east of Lvliang Mountain, Xinghuacun Town is under the jurisdiction of Fenyang, Shanxi Province. The long history of Baijiu-making in Xinghuacun Town has been proved by the drinking vessels in the periods of the Yangshao, Longshan, Xia and Shang excavated there, among which the small pointed bottom urn is the most typical. The widespread knowledge of Baijiu-making techniques in Xinghuacun Town has a close connection with two historical events. One is that because of the immigration and land reclamation policies from the 2nd year of Hongwu, Ming Dynasty, a large number of Baijiu-making masters migrated to other places in the country. The other is that with the rise of merchants from Shanxi in the middle and late Ming Dynasty, Baijiu brands from Xinghuacun Town were sold all over the country, which brought prosperity to the Baijiu-making industry. Xinghuacun Town is now the leading producer of the Light flavor type Baijiu and the producing base of the Fenjiu Xinghuacun Group.

Located on the west bank of Fujiang River, one of the tributaries of the Yangtze River, Tuopai Town is under the jurisdiction of Suining, Sichuan Province. According to the *Records of Sichuan*, the origin of the Tuopai Qu Baijiu production can date back to the Tang Dynasty when it was named 'Shehongchun Baijiu'. The line 'the shehongchun Baijiu is still green in the winter' from Fu Du is praise for it. In the 35th year of the Republic of China (1946), Tianqu Ma, a winner of the imperial exams at the provincial level changed the name of 'Shehongchun Baijiu' into 'Tuopai Qu Baijiu' based on the words in an arch 'tuo spring brews excellent Baijiu; the brand will be praised for a long time'. Tuopai Town is the leading producer of Strong flavor Baijiu and the producing base of the Shede Group.

Located in the south of Huanghuihai Plateau, Gujing Town, called Jiandianji in the past, is under the jurisdiction of Bo County, Anhui Province. In the late Eastern Han Dynasty, Cao Cao presented 'Jiuyunchun Baijiu' made in Jiandianji to the Emperor Xian of the Han Dynasty. During the Wanli era of the Ming Dynasty, li Shen, an elder official, presented 'Jiandianji Baijiu' to the Emperor Wanli. Therefore, the major feature of the Baijiu made in Gujing Town is to use it as a tribute. So far, the biggest industrial cluster of Baijiu in Anhui Province is located in Gujing Town. Among more than 100 Baijiu enterprises, Gujinggong Group is the leading one.

Located in the central Shandong Peninsula, Jingzhi Town is under the jurisdiction of Anqiu, Shandong Province. Jingzhi Town, Yanshen Town and Zhangqiu Town are the 'Three Ancient Towns' in Shandong Province. According to *The Records of Anqiu*, in the Hongwu era of the Ming Dynasty, the Baijiu-making industry developed in Jingzhi Town. During the Qing Dynasty, the Baijiu-making industry in Jingzhi Town was flourishing. The Baijiu made in Jingzhi Town was the mellowest one in Anqiu according to *The Records of Anqiu* in the Guangxu era. It was recorded in *The General Records of Shandong* that Baijiu is flourishing in Jingzhi Town, Anqiu. Jingzhi Town was awarded 'The First Town of Chinese Sesame Flavor Type Baijiu' by the Chinese Light Industry Council and the Chinese Alcoholic Drinks Association in 2012. Jingzhi Group is the leading enterprise in Anqiu.

39. The Story of Erguotou Baijiu

Erguotou, a Baijiu of strong Beijing characteristic, is the only Baijiu named after its manufacturing technique and is popular all over China, especially the Beijing area, for its refreshing taste. The representative Erguotou Baijiu products are shown in Figure 3.13 (Red Star Erguotou Baijiu) and Figure 3.14 (Niulanshan Erguotou Baijiu).

Figure 3.13. Red Star Erguotou Baijiu products from Beijing.

Figure 3.14. Niulanshan Erguotou Baijiu products from Beijing.

Baijiu-making in Beijing originated from the Yuan Dynasty and thrived during the Qing Dynasty. The Baijiu vapors from distillation are condensed into liquids at the bottom of the cooler by cold water, and the liquids go into the Baijiu container through the inducing tubes. The traditional manufacturing technique of Erguotou originated from the unique skills of 'cutting out both ends of the distillate and taking the middle' invented by three brothers, Cunren Zhao, Cunyi Zhao and Cunli Zhao, from the Baijiu brewing workshop 'Yuanshenghao' located in Qianmenwai, Beijing, in the 19th year of Emperor Kangxi (1680) during the Qing Dynasty. During distillation, the cold water in the pot was controlled according the size of the distillate foam, and the liquids condensed from different coolers were collected separately. The liquid from the first cooler is called 'the initial distillate' that tastes acrid because of the impurities with low boiling points. The liquid from the third cooler is called 'the last distillate' that also does not taste good because of low alcohol content and the presence of high boiling point components. The liquid from the second cooler tastes refreshing with less impurities and appropriate alcohol content. The traditional Erguotou comes from the distillate of the second cooler with the best quality, and thus it is called Erguotou. Nowadays, Baijiu distillation in domestic Baijiu factories follows the manufacturing technique of Erguotou by obeying the rules of 'cutting out both ends of the distillate' and 'picking liquids according to the quality', which are in line with the essence of the Erguotou Baijiu steaming technique.

The leading brands of Erguotou Baijiu include Red Star Erguotou, Niulanshan Erguotou and Huadu Erguotou.

40. Shibajiufang Baijiu

Stemming from the former name of the distillery 'Shibajiufang' (18 Baijiu workshops), Shibajiufang (SBJF) Baijiu, as shown in Figure 3.15, is one of the two famous Baijiu brands of Hebei Hengshui Laobaigan Liquor Co., Ltd. 'SBJF' is the collective name

Figure 3.15. Shibajiufang Baijiu from Hebei Province.

of the 18 Baijiu brewing workshops producing Hengshui Laobaigan Baijiu on both sides of the Fuyang River, Hengshui.

Hengshui, called Tao Town or Tao City in the past, has a long history of Baijiu-making. It was said that 'Tao City is not big, but has 18 Baijiu-making workshops'. Till the Mid-Qing Dynasty, Hengshui had been the famous Baijiu-making center all over the country. Most of the '18 Baijiu-making workshops' were established during the Qing Dynasty and some during the Ming Dynasty. The manufacturing techniques of the workshops originated from the same original protocol, and were similar but different. According to documents from the 1930s, the 18 Baijiu-making workshops referred to Deju, Guangju, Tiancheng, Xinda, Dechang and Jixing located in Tongshang Street, Fuxinglong, Xingyuanxiang and Hengjucheng located in Muchang Street, Chengxinghao and Qingshezeng located in Bizishi, Hengdecheng and Tianfenghao in Caishi Street, Yixinglong and Fujuxing in Wenjin Street, and Hengshenghao, Yuanshenghao and Deyuanyong in Hexi Street.

Hengshui was liberated on December 16th, 1945. In the spring of 1946, following instructions from the Ji'nan Administrative office, the government of Hengshui Town bought the 18 privately owned Baijiu-making workshops and established the Hengshui Baijiu Distillery of Ji'nan (predecessor of Hengshui Laobaigan Liquor Co., Ltd), the first state-owned distillery before the foundation of the People's Republic of China. With 70 years of development, it has been the largest, leading enterprise of the Laobaigan flavor type Baijiu in China.

41. Alcohol Vessels and China

Alcohol vessels are used to serve alcohol. The development and evolution of alcohol vessels is closely related to the level of productivity and the level of science and technology. It also reflects the values and aesthetic standards of the population. The unique culture of alcohol vessels has formed gradually.

Based on the materials, the development of Chinese alcohol vessels has gone through periods of pottery, bronze, porcelain, glass and so on. The earliest alcohol vessels were made of pottery with a history of over 9000 years. Some alcohol vessels made of pottery are still used today. For example, Jingzhi Distillery still keeps the former peach blossom earthen jar (Weng), which is used in Baijiu brewing when peach trees are in full bloom every year. The bowls used to drink Baijiu by people of the Tujia nationality are made of pottery in Enshi, Hubei Province.

Alcohol vessels made of bronze such as jue, jiao, gu, zhi, jia, zun, hu, you, fangyi, dou, shao and jin reached a peak during the Shang Dynasty, but now they are of little use other than as works of art.

Porcelain is the invention of the Chinese and is made based on pottery-making technology. The white pottery in the Shang Dynasty laid the foundation for ancient porcelain. Blue-glazed porcelain from sites of the Shang Dynasty and Western Zhou Dynasty is recognized as the ancient porcelain. The porcelain from the Eastern Han Dynasty to the Wei-Jin Period is mostly celadon porcelain. White-glazed porcelain originated in the Northern and Southern Dynasties and flourished during the Sui Dynasty. Porcelain in the Tang Dynasty had almost reached today's standards of the advanced fine porcelain. Porcelain-making technology had completely reached maturity during the Song Dynasty.

The development of porcelain not only generated the evolution of alcohol vessels but also brought about the prosperity of the brewing industry in China. Even now, Chinese alcohol vessels made of porcelain are widely used in the process of producing, aging, selling and drinking of Chinese Baijiu and Huangjiu. Fen Baijiu, the ancestor of Chinese Baijiu, is fermented in Digang, porcelain jars buried underground. Hengshui Laobaigan Baijiu is also fermented in Digang.

Research has shown that the air permeability of porcelain improves Baijiu flavor during the aging process. The famous Huangjiu brands such as 'Nverhong' and 'Zhuangyuanhong' are made by putting the fresh Huangjiu into porcelain jars and keeping them underground for about 18 years. Nowadays, porcelain is preferred for the aging of all kinds of famous and excellent Baijiu in China.

Coincidently, the English word 'china' stands for both porcelain and the country China. Chinese people use porcelain jars to store Baijiu and Huangjiu, porcelain bottles to contain Baijiu and Huangjiu (shown in Figure 3.16) and porcelain cups to drink Baijiu and

Figure 3.16. Porcelain bottles to contain Huangjiu (left) and Baijiu (right).

Figure 3.17. Porcelain cups to drink Baijiu.

Huangjiu (shown in Figure 3.17). The dishes that go with Baijiu are put in plates and trays made of porcelain, and the bowls to contain rice are made of porcelain, too. This is typical of Chinese culture.

42. The Tasting and Appraisal of Baijiu

In ancient times, the saying was 'singing while enjoying alcohol if you still can', and today it is, 'a pot of Baijiu would comfort all my efforts for life'. In daily life, the taste evaluation of Baijiu is not simply a skill

Figure 3.18. The scene of tasting and appraisal of Baijiu.

but also a science that brings pleasure and health. Figure 3.18 shows a scence of tasting and appraisal of Shede Baijiu from Sichuan Province.

The color is observed by the eyes. A glass is filled 1/3–3/5 (v/v) with Baijiu, and its color, transparency and the presence of suspended particles or sediment are examined by observing from different angles, such as horizontally and vertically. Then, the cup-hanging is examined by shaking the glass gently. The cup-hanging is easy to be noted for a quality Baijiu as a silky liquid which is clear and transparent with no impurities.

The aroma is examined by smelling. The distance between your nose and the glass should be kept at 1–3 cm when smelling a Baijiu sample, and inhaling without exhaling toward it. One shall also shake the glass slightly in front of the nose to enjoy the overflowing aroma. Baijiu may have grainy, flowery and sweet aroma, a wonderful pleasant feeling.

The taste is examined by drinking. The Baijiu intake should be slow and stable, to cover the entire tongue, since the tip and edge of the tongue are sensitive to salty taste. The front part of the tongue is more sensitive to sweetness, and both sides of the tongue near the jaws are sensitive to sourness. The rear part of the tongue is sensitive to bitter and spicy tastes. It is important to keep the liquid in the mouth, to slowly taste and swallow, and to enjoy the aftertaste following the sweet and strong flavor. It is just like life, rich in flavor and taste with no words to fully describe it.

No matter how wonderful a Baijiu tastes, one should not drink too much. *Dietary Guidelines for Americans from 2015 to 2020* published by the U.S. Department of Health and Human services and the U.S. Department of Agriculture indicates that women should not have more than 1 serving unit each time and men should not have more than 2 serving units each time. In a single day, a woman should not drink more than 3 serving units and a man should not drink more than 4 serving units. 1 serving unit is equal to liquor with ethanol of 0.6 ounce (almost 18 mL). In general, 1 serving unit is equal to 355 mL of 5° beer or 150 mL of 12° grape wine or 45 mL of 40° Baijiu or 30 mL of 60° Baijiu.

As the saying goes, 'drinking is best to be ended at a slight drunkenness whereas flowers are most beautiful at halfway blossoming'. Human beings must stick to the principle that moderate drinking at an appropriate time with excellent manners is necessary. One of the authors of this book said in his poem *Healthy Drinking* that '100 mL of Baijiu promotes communications; 250 mL of Baijiu makes one dare to brag; more than 250 mL of Baijiu would make one embarrassing; so moderate drinking is necessary for one's health'.

43. The Drinking Traditions of Huangjiu

Drinking tea follows its ways and so does alcohol. Huangjiu is mild in nature and drinking it slowly brings the best flavor.

In the seasons when the temperature is below 10°C, Huangjiu is usually warmed before drinking. The way to warm Huangjiu in Shaoxing is 'Chuantong in the boiled water', and it is very interesting. Pour Huangjiu into Chuantong and then put it into the boiled water. The temperature of Huangjiu rises gradually and it is ready for serving with a mild taste when it is aromatic. It is important not to overheat it. Pour Huangjiu back into the bottle(s) after heating and then serve in the cups. The amber liquid ripples in the cup with a pleasant aroma. The warm Huangjiu warms one's stomach and invigorates the circulation of blood. The strength of it disappears fast and makes one feel more comfortable. Mr. Yuanpei Cai, an educator in modern China, used to warm Huangjiu in thermos bottles at home and enjoy it with friends. They often drank 200 mL per person each meal and never got drunk.

In the middle of summer, it is suitable to store Huangjiu in the fridge at the temperature about 3°C and then drink it directly or with ice cubes. The mixture of amber Huangjiu and glistening ice cubes looks enchanting and tastes refreshing, and it is not easy to get drunk.

The most graphic verb to describe Huangjiu drinking is 'mi', the combination of smelling, sipping and tasting, which makes Huangjiu tastier and brings pleasure while closing one's eyes. 'Mi' helps one taste the various flavor elements that are released gradually into the nose and mouth while drinking Huangjiu, with a rich after-flavor and aftertaste.

Huangjiu is suitable for slow drinking and the best dishes to eat with it are chewy and rich in flavor such as flavored boiled peanuts, fennel beans and dried tofu. Huangjiu is perfectly matched with Chinese mitten crabs as the saying goes, 'eating the crabs while

Figure 3.19. The drinking traditions of Huangjiu. (The hand drawing is courtesy of Zimei Zhao, BTBU.)

drinking Huangjiu'. Chinese mitten crabs are tasty but 'cold' in nature, while Huangjiu is warm in nature according to traditional Chinese medicine, so there is less health concern if eating crabs with Huangjiu. The fishy smell is also neutralized by Huangjiu, so that the crabs taste more delicious. The drinking traditions of Huangjiu can be expressed vividly by the hand-drawn in Figure 3.19.

It is a great pleasure to drink Huangjiu slowly with several nice dishes together with a few intimate friends. The slight drunkenness creates a relaxing atmosphere, the desire to confide to each other, the impulse to express feelings, the courage to know more about each other, the expectations to explore, which shortens the distance from each other, and opens one's mind inch by inch.

Chapter 4
Brewing

44. Brewing of Strong Flavor Type Baijiu

The Strong (Nong) flavor type Baijiu was initially called Luzhou-flavor type Baijiu. It was named after the characteristics of Luzhou Laojiao Baijiu, and has been known as the Strong flavor type Baijiu since the 1980s. Strong flavor type Baijiu has the characteristics of an enriched strong flavor, a soft sweet combined with a mild hint of astringent taste, harmonious and balanced aroma, a sweet hint at the beginning, a smooth and clear mouthfeel, and long-lasting after-flavor and aftertaste.

According to the national standard, the Strong flavor type Baijiu is defined as a spirit product that is made from grains by traditional solid-state fermentation, followed by distillation, aging and blending, without adding edible alcohol or flavor substances produced from non-alcoholic fermentation, and has ethyl hexanoate as the primary flavor component.

The Strong flavor type Baijiu can be brewed using sorghum, rice, glutinous rice, wheat and maize as the starting materials, Jiuqu (medium-temperature Daqu or Fuqu) as a saccharifying and fermenting starter and a solid-state fermentation process characterized by 'Hunzheng Xucha (HZ-XCA)'. 'HZ-XCA' means that the mixture of the crushed new grain ingredients is mixed with the Jiupei from the last fermentation and subjected to a steam distillation process in a steamer barrel called Zengtong. This operation is also called 'Hunzheng Hunshao (HZ-HS, steaming the starting materials and the fermented grains together)'. After removing from Zengtong, the mixture is cooled, Qiu is added and it is fermented in the pit again. The above process is operated repeatedly. Most of the Strong flavor type Baijiu is produced by this process. The characteristics of this brewing process are that the flavor substances contained in several different grains, such as esters, phenols and/or vanillin, can be brought into Baijiu during the mixing and steaming process, which enhances the flavor of the Baijiu. These aromas are called 'grain aroma', such as sorghum aroma. In the mixing and steaming process, the acid and water contained in the Jiupei accelerate the gelatinization of starch in the grain materials and facilitate the fermentation process. Addition of new grains can reduce the amount of auxiliary materials (rice husk, sorghum husk, etc.) and improve the quality of Baijiu. The starting grains can be fermented more than three times, so the utilization of raw materials is efficient. The fermented grains and distillery grains are collectively called distillers' grains by Southern distilleries. Because the fermented grains can be recirculated and fermented many times, it seems that they would never be discarded, so people often regard this kind of grain as 'millennial distillers' grains'. The longer the fermentation time of the Jiupei, the more the preconditions are accumulated, which play important roles in enhancing the quality and flavor of Baijiu. The common saying 'a thousand years of cellar and ten thousand years of distillers' grain' fully shows how the quality of the Strong flavor type Baijiu is closely related to the cellar and distillers' grains. The cellar used in the brewing process of Luzhou Laojiao Baijiu, which is a typical representative of the Strong flavor type Baijiu, is shown in Figure 4.1.

The Strong flavor type Baijiu is a precious historical and cultural heritage of China, unique in the world, and is currently has the largest production and sales. More than 70% of Chinese Baijiu products are of the Strong flavor type Baijiu.

According to the starting grains, the Strong flavor type Baijiu can be divided into two types, the single-grain and the multi-grain Baijiu. The Strong flavor type Baijiu with single grain is brewed with

Figure 4.1. The cellar used in the brewing process of Strong flavor type Baijiu.

sorghum, a single raw material, and the typical representative is Luzhou Laojiao Baijiu from Sichuan Province. The Strong flavor type Baijiu with multi-grain is brewed from a mixture of wheat, rice, corn, sorghum and glutinous rice, and is represented by Wuliangye Baijiu from Sichuan Province. Sorghum is the dominant raw material for Baijiu-making, followed by rice, wheat, glutinous rice, corn, barley, highland barley and so on. The following properties are generally required for grains for Baijiu production: high carbohydrate content, appropriate protein and tannin contents, suitable for absorption and utilization by microbes, ease of storage, strictly controlled water content in order to prevent mildew and decay.

The Strong flavor type Baijiu can also be divided into Daqu and Fuqu flavor types according to the different saccharifying and fermenting starters. Southern China is dominated by the Daqu Strong flavor type, in areas such as Sichuan, Anhui and Jiangsu Provinces. The representative brands are Wuliangye Baijiu, Luzhou Laojiao Baijiu, Yanghe Daqu Baijiu, Jiannanchun Baijiu, Gujinggong Baijiu, Quanxing Daqu Baijiu and Shuanggou Daqu Baijiu. Daqu strong flavor type Baijiu production areas in northern China include Bancheng Shaoguo Baijiu in Chengde, Hebei Province, and Taoerhe Baijiu in Baicheng, Jilin Province. In northern China, the Fuqu Strong flavor type Baijiu is the primary flavor type, in areas such as Liaoning, Jilin, Inner Mongolia and Hebei Provinces, whose representative brands are Jinzhou Qu Baijiu of Liaoning, Dehui Daqu Baijiu and Longquanchun Baijiu of Jilin, and Ningcheng cellar Baijiu and Chifeng Chenqu of Inner Mongolia. In addition, many distilleries in Shandong Province use Daqu and Fuqu together to produce the Strong flavor type Baijiu, such as Bandaojing, Jingzhi and a few other brands.

45. Brewing of Light Flavor Type Baijiu

The Light (Mild) flavor type Baijiu has a long history, unique technology, mellow and sweet taste, pure aroma and is very popular in northern and parts of southern China. The Light flavor type Baijiu is also the best base alcohol for producing liqueur and medicinal spirits due to its pure aroma.

According to the national standard, the Light flavor type Baijiu is defined as a kind of spirit made from grains through a traditional solid-state fermentation process, distillation, aging and blending, without adding edible alcohol or flavor substances produced from non-alcoholic fermentation, and has ethyl acetate as the primary flavor component.

The Light flavor type Baijiu is brewed with sorghum and other grains as the starting materials, and Jiuqu (medium-temperature Daqu, low-temperature Daqu, Xiaoqu or Fuqu) is used as the saccharifying and fermenting starter. The solid-state brewing process is named 'Qingzheng QCA (QZ-QCA)'. The 'QZ-QCA' means that the starting materials and auxiliary materials are steamed separately, and mixed proportionally, and then the starter is added to the mixture for the first fermentation. After the fermented grains are steam-distilled, no new grains are added, but the starter is added before the second fermentation. Finally, the second fermented grains are steamed and the residue is discarded. The auxiliary materials used in the abovementioned process may be rice husk, rice bran, sorghum husk, corn cob, fresh distillers' grains and peanut peel. They can adjust the concentration of starch in grains, reduce or improve acidity, absorb the alcohol, maintain the content of Huangshui (the yellow serofluid produced during fermentation process), maintain the certain porosity and oxygen content of Jiupei, and increase interfacial effects, so as to promote the conversion of starch to sugar in the process of steaming grains. Sugarization also ensures that the fermentation and distillation of Baijiu go smoothly. The brewing technique of the Light flavor type Baijiu emphasizes 'steaming separately, discharging the impurities, sanitation, and hygiene'. The critical point is to ensure the clarity and purity from the beginning to the end of the process, including steaming raw materials and auxiliary materials separately, a 'clear and pure' fermentation and a 'clear and pure' distillation. The advantages of this brewing process are a short fermentation period, a low production cost, a high Baijiu yield from the raw materials, reducing grain utilization, using Digang and a low temperature to ferment, a well-maintained processing facility and relatively fewer components in the Baijiu. Figure 4.2 shows the Digang used in the brewing process of Fen Baijiu, a typical representative of the Light flavor type Baijiu.

Figure 4.2. Digang used in the brewing process of Light flavor type Baijiu.

The Light flavor type Baijiu products are produced in many areas of China, and have the most varieties of flavor types. According to the difference of Jiuqu, the Light flavor type Baijiu can be Daqu Light flavor type Baijiu, Xiaoqu Light flavor type Baijiu and Fuqu Light flavor type Baijiu. Xiaoqu flavor is dominant in southern China, while Daqu and Fuqu flavors are dominant in northern China.

One of the typical representatives for the Daqu Light flavor type Baijiu is Fen Baijiu of the Shanxi Xinghuacun Fen Liquor Factory Co., Ltd. It is made using sorghum as the raw material and Daqu as the saccharifying and fermenting starter. The commercial product has the characteristics of 'soft mouthfeel at the beginning, sweet taste note, pure light aroma and flavor, and long-lasting after-flavor'. Huanghelou Baijiu produced in Wuhan Tianlong Huanghelou Liquor Co., Ltd., Guose Light Baijiu in Henan Baofeng Liquor Co., Ltd., and Erguotou Baijiu of Beijing are all Daqu Light flavor types of Baijiu with their own unique brewing techniques.

The Xiaoqu Light flavor type Baijiu is produced by a solid-state fermentation of sorghum, corn and rice as raw materials and Xiaoqu as the saccharifying and fermenting starter. Its production has a long history and it is usually made in a large amount. It is popular in Sichuan Province, Chongqing City, Yunnan Province, Guizhou Province, Hubei Province and a few other locations. The products have their own unique styles. Typical representatives are Jiangjin Laobaigan Baijiu and Yongchuan Sorghum Baijiu from Chongqing City, Yulin Baijiu from Yunnan Province, and Jinpai Xiaoqu Baijiu from Hubei Province.

The Fuqu Light flavor type Baijiu is produced using a solid-state fermentation technique with sorghum and maize as raw materials, and Fuqu as a saccharifying and fermenting agent. The process is short in a production cycle time and high in Baijiu yield. The typical representatives of this type of Baijiu are Liuquxiang Baijiu from Shanxi Province, Lingta Baijiu from Liaoning Province, Laobaigan Baijiu from Harbin City and Caoyuanwang Baijiu from the Inner Mongolia Autonomous region.

46. Brewing of Sauce Flavor Type Baijiu

The Sauce (Jiang) flavor type of Baijiu, also known as the Moutai flavor type Baijiu, is well known for its elegant and graceful aroma and flavor, delicacy, mellow, rich and full body, and lasting after-aroma, and is deeply loved by consumers.

In the national standard, the Sauce flavor type Baijiu is defined as a kind of Baijiu with a Sauce flavor note, which is made from sorghum, wheat and water through a traditional solid-state fermentation, distillation, aging and blending without adding edible alcohol or flavor substances produced from non-alcoholic fermentation.

The Sauce flavor type Baijiu can be divided into Daqu and Fuqu Sauce flavor types according to the different saccharifying and fermenting starters.

Moutai Baijiu in Guizhou Province and Lang Baijiu in Sichuan Province are typical representatives of the Daqu Sauce flavor type Baijiu. The brewing technology of the Daqu Sauce flavor type Baijiu is very distinctive. Sorghum is used as the starting material, and high-temperature Daqu made using wheat is used as the saccharifying and fermenting starter. The process includes raw material feeding for two times, repeating the fermentation for eight runs and Baijiu picking for seven times. Then, three typical Baijiu bases with sauce, mellow and cellar flavors are blended with the different rounds of fermented Baijiu after each has had a long storage time. 'Twice of feeding' procedure refers to the process in which the raw material, sorghum, is first put in with about half of the total volume, followed by steaming, adding starter and fermentation. The remaining half is added to the fermented grains after the first fermentation, and the fermentation is continued after distillation and adding starter. 'Eight-round fermentation' refers to the process of discarding the distillers' grains after eight times of repeated fermentation and distillation from the first feeding. After the first and second rounds of adding raw materials, from the third round, only Daqu is added. 'Seven times of Baijiu picking' refers to the distilled Baijiu that is collected after the last seven times of fermentation and stored separately according to the aroma characteristics and rounds, whereas the distillation of Baijiu after the first fermentation is directly added back to the fermented grains for continued fermentation.

The technological characteristics of the Daqu Sauce flavor type Baijiu can be summarized as 'four high, two long, one big, and one adequate'. The 'four high' is as follows: (1) high-temperature Daqu is used as a saccharifying and fermenting starter, and it is prepared at a temperature above 60°C; (2) high-temperature accumulation means that the fermented grains are kept in the air to reach 45–50°C before being moved into pits for further fermentation; (3) high-temperature fermentation, that is, the temperature in the fermentation pits can reach 42–45°C; and (4) high-temperature distillation means that the distillation temperature of the Sauce flavor type Baijiu is higher than that of the other flavor types of Baijiu. One of the 'two long' refers to a long production cycle. The production cycle is about one year from the beginning of feeding the grains to the end of Baijiu production. The other is a long storage period, generally more than three years. Long-term storage is important to ensure the stability of flavor quality of the Sauce flavor type Baijiu. Figure 4.3 shows the long-term storage of Lang Baijiu. 'One big' refers to the large amount of starter used in the Sauce flavor type Baijiu brewing process, which is the largest amount of starter used in all flavor types of Baijiu, and the ratio of starter to raw materials is 1: (0.85–0.95). 'One adequate' refers to

Figure 4.3. The long-term storage of Sauce flavor type Baijiu.

the multiple cycles of fermentation and distillation, usually eight fermentation and seven distillation rounds.

The Fuqu Sauce flavor type Baijiu was developed on the basis of imitating Moutai Baijiu in the 1950s. It was produced by using Fuqu as a saccharifying and fermenting starter and using the 'Qingzheng-XCA (QZ-XCA)' brewing technique, but while retaining the technological characteristics of high-temperature accumulation, fermentation and distillation. The so-called 'QZ-XCA' refers to the process during which the separately steamed starting grains are mixed with the remaining fermented grains from the former fermentation cycle before adding a Fuqu starter for further fermentation. The typical representatives of the Fuqu Sauce flavor type Baijiu are Yingchun Baijiu of Hebei Langfang Yingchun Liquor Co., Ltd., Lingchuan Baijiu of Liaoning Jinzhou Lingchuan Liquor Factory and Qianchun Baijiu of Guizhou Qianchun Liquor Co., Ltd. The Fuqu Sauce flavor type Baijiu has the advantages of high yield, and short fermentation and storage periods.

47. Brewing of Rice Flavor Type Baijiu

The Rice (Mi) flavor type Baijiu is a kind of Xiaoqu Baijiu, which is derived from rice alcohol and Huangjiu. It has the characteristics of being colorless, clear and transparent, with an elegant and graceful honey flavor, sweet beginning taste, refreshing and long-lasting pleasant taste.

The Rice flavor type Baijiu is defined by the national standard as a kind of spirit made from rice and other grains by traditional semi-solid fermentation, distillation, aging and blending, without adding edible alcohol or flavor substances produced from non-alcoholic fermentation, and it has the primary mixed flavor of ethyl lactate and beta-phenylethanol.

The Rice flavor type Baijiu is made with rice as the starting grain, Xiaoqu as a saccharifying and fermenting starter, solid-state fermentation at the early stage followed by liquid-state fermentation at the later stage and liquid-state still distillation. Due to the solid-state saccharification and bacterial culture process, the amount of Jiuqu used is very small, generally 0.8–1.0% of the grains, which is one of the major differences from the Baijiu production process for other flavor types. Since the 1980s, the steaming, saccharification, fermentation and distillation processes of the Rice flavor type Baijiu have been mechanized, which reduces labor intensity, improves production efficiency and stabilizes the product quality.

The essential characteristic of Rice flavor type Baijiu production is the semi-solid fermentation technique, which is a method between traditional solid-state and modern liquid-state fermentation techniques, and is one of the biggest differences compared with other flavor types of Baijiu. According to the state of materials during the fermentation and distillation processes, the brewing techniques of Chinese Baijiu can be divided into the solid, semi-solid and liquid methods. The solid-state brewing method is the traditional technique for Baijiu production in China, with a high quality and good flavor, but a low yield and high labor intensity. The liquid-state brewing process is a new technique which appeared in the 1960s, whose main advantages are high mechanization, low labor intensity, high Baijiu yield and high production efficiency. However, the flavor of Baijiu using the liquid-state brewing method is inferior to that of the solid-state brewing process. The semi-solid process is between the above two processes, and it is also a traditional brewing process for Chinese Baijiu.

The main producing areas of the Rice flavor type Baijiu are in Guangxi, Guangdong, Hunan, Hubei, Jiangxi, Fujian, Guizhou, Yunnan and Sichuan Provinces. Among them, typical representatives are Guilin Xiangshan and Sanhua Baijiu in Guangxi Province, Liuyang Xiaoqu Baijiu in Hunan Province, and Changle Baijiu and Conghua Sanhua Baijiu in Guangdong Province.

48. Brewing of Feng Flavor Type Baijiu

The Feng flavor type Baijiu has a long history in China, and it was first established in 1993. It has the characteristics of the Light and Strong flavor types of Baijiu with the characteristics of mellow and rich flavor, sweet, refreshing and harmonious flavors with a long-lasting after-flavor.

The Feng flavor type Baijiu is defined by the national standard as the spirit with composite flavor of ethyl acetate and ethyl hexanoate, without adding edible alcohol or flavor substances produced from non-alcoholic fermentation, and is made from grains by the traditional solid-state fermentation, distillation, aging by Jiuhai and blending.

The Feng flavor type Baijiu is brewed with Daqu as a saccharifying and fermenting starter and sorghum as the starting material by the 'QZ-XCA' technique as mentioned in the section titled 'Brewing of Sauce Flavor type Baijiu'. The technological characteristics are as follows: First, Daqu is made from barley and pea instead of wheat, and is a medium- to high-temperature Daqu. Second, the fermentation period is short, which was 11–14 days earlier, and was extended to 18–23 days, and is the shortest fermentation period among national famous Baijiu brands. Third, the fermentation is carried out in new mud pits, and the pits need to be re-mudded every year. Fourth, it is stored in Jiuhai, different from other types of Baijiu, which are usually porcelain jars. The fresh Baijiu needs to be stored for three years before it can be used to blend and produce commercial Baijiu. At present, the brewing technique of the Feng flavor type Baijiu has been improved in several ways. First, wheat is added to the raw materials of Daqu production; second, the fermentation time is extended to 30 days; third, the flavor types have been developed toward Feng-Strong, Feng-Strong-Sauce and other multiple flavor types.

The Feng flavor type Baijiu is mainly produced in northwest and northeast China. Typical representatives are Xifeng Baijiu of Shanxi Xifeng Liquor Co., Ltd. and Taibai Baijiu of Shanxi Taibai Liquor Co., Ltd.

49. Brewing of Mixed Flavor Type Baijiu

The Mixed (Jian or Nongjiang) Flavor type Baijiu was developed in the early 1970s by innovatively combining the techniques of the Strong and Sauce flavor types of Baijiu on the basis of the production experience of famous Baijiu. At present, the Mixed Flavor type Baijiu is dominated by the Strong-Sauce mixed flavor type, which has the elegance and delicacy of a sauce flavor type Baijiu and the sweetness and refreshment of a Strong flavor type Baijiu, with comfortable taste and unique style.

According to the national standard, the Mixed Flavor type Baijiu is defined as the spirit with a unique Strong-Sauce mixed flavor, which is made from grains by traditional solid-state fermentation, distillation, aging and blending without adding edible alcohol or flavor substances produced from non-alcoholic fermentation.

Baiyunbian Baijiu of Hubei Baiyunbian Liquor Co., Ltd. and Yuquan Baijiu of Heilongjiang Yuquan Liquor Co., Ltd. are typical representatives of the Strong-Sauce mixed Flavor type Baijiu. The flavor characteristics of these two kinds of Baijiu are different because of their different preparation techniques.

Baiyunbian Baijiu from Hubei Province is characterized by a strong flavor combined in a sauce flavor. It is made using sorghum as the starting material, and combined with the brewing technique of the Strong flavor type Baijiu on the basis of the process of the Sauce flavor type Baijiu. In general, the procedure is as follows: 3 to 4 feeding times, 6 stacking times, steaming, a combination of 'Qingzheng Qingshao' (QZ-QS, means steaming raw grains and Jiupei separately) and 'HZ-HS', cellar fermentation, nine rounds of the procedure, 7 times of picking up Baijiu and long-term storage and blending. Compared with the brewing technique of the Sauce flavor type Baijiu, the technology used in making Baiyunbian Baijiu has one more round of fermentation and steaming process, a total of 9 rounds. Besides the first and second rounds of feeding as the Sauce flavor type Baijiu, 1–2 times of feeding in the seventh and eighth rounds are needed, totally 3–4 times of feeding. Additionally, high-temperature Daqu and medium-temperature Daqu are used in combination, a high-temperature Daqu fermentation during the first several rounds and a medium-temperature Daqu during the latter 2–3 rounds or the eighth round. High-temperature accumulation and high-temperature fermentation are used in the first few rounds, followed by 2–3 rounds or eight rounds of the technique of the Strong flavor type Baijiu with low-temperature fermentation.

The characteristic of Yuquan Baijiu of Heilongjiang Province is that the strong flavor contains sauce flavor, which is blended with the Sauce and Strong flavor types of Baijiu according to a certain proportion. The basic Baijiu of the above two flavor types is produced by different brewing techniques, and then used after aging separately. This process is also known as 'type-dependent fermentation, type-dependent aging, and scientific blending'. The Strong flavor type Baijiu is produced with sorghum as the starting material and a medium-temperature Daqu as saccharifying and fermenting starter through the solid-state brewing technology of 'HZ-XCA'. The Sauce flavor type Baijiu is produced by the typical technique of one feeding and six rounds of fermented grains. The grains after six rounds of fermentation are continued to be used in accordance with the 'HZ-HS' process.

50. Brewing of Dong Flavor Type Baijiu

The Dong flavor type Baijiu, also known as the Herblike flavor type Baijiu, is a new genre and flavor type of Chinese Baijiu. The body of Baijiu is clear and transparent with elegant, comfortable, medicinal and mellow flavor, and long-lasting aftertaste. 'Ester flavor, alcohol flavor and herbal flavor' are three important characteristics of the Dong flavor type Baijiu.

The Dong flavor type Baijiu is defined by the local standard of Guizhou Province as a type of spirit with the grain materials of sorghum, wheat and rice. Daqu and Xiaoqu are made by unique traditional techniques, and are fermented in big and small cellars by a solid-state method, then distilled together, and finally stored for a long time and blended without adding edible alcohol or flavor substances produced from non-alcoholic fermentation.

The Dong flavor type Baijiu is brewed from sorghum by solid-state fermentation of Xiaoqu (also known as a rice starter) prepared from rice, and Daqu by solid-state fermentation of wheat to produce flavor-fermented grains (one kind of fermented grain). Then, the fermented grains are placed below the flavor-fermented grains for solid-state distillation. The technological characteristics are that Daqu and Xiaoqu are used simultaneously, and in the process of preparing Daqu and Xiaoqu, Chinese herbal medicines are added; more than 130 kinds of Chinese herbal medicines are used altogether. The second is the crossing-steaming of double fermented grains which is the process of distilling the two kinds of fermented grains together. This kind of flavor crossing distillation process has an important impact on the development of the Baijiu industry and has been widely used in the production of many types of Baijiu.

The typical representative of the Dong flavor type Baijiu is Dong Baijiu of Guizhou Dong Jiu Co., Ltd. There are similar flavor types of Baijiu produced in Sichuan, Jiangxi, Shandong, Hubei, Yunnan and Henan Provinces.

51. Brewing of Chi Flavor Type Baijiu

The Chi flavor type Baijiu, also known as fermented soybean-flavored Yubingshao Baijiu and Roubingshao Baijiu, is a kind of Xiaoqu Baijiu. It has the characteristics of being clear and transparent, colorless to light yellow, crystal pure, unique fermented soybean flavor, mellow and smooth, and a refreshing aftertaste.

The national standard defines the Chi flavor type Baijiu as a kind of spirit with fermented soybean characteristics, which is made using rice or pre-crushed rice as the starting material, and cooking, using large Jiuqu cakes as the primary saccharifying and fermenting starter, fermenting while saccharifying, distilling, aging with stale flesh and blending, without adding edible alcohol or non-self-fermenting coloring, aroma and flavor substances.

The Chi flavor type Baijiu was separated from the Rice flavor type Baijiu in 1984, and was defined as a new flavor type, and its brewing technique was a typical representative of simultaneous fermentation and saccharifying processes. Because there is no separated saccharification process, the amount of Jiuqu used is larger than that for the Rice flavor type Baijiu (18–20% of the grains), and it uses a traditional liquid-state fermentation process. 'Aged meat immersion', keeping a piece of fatty pork in the freshly distilled Baijiu for a predetermined period of time, is another unique processing step in the brewing process for the Chi flavor type Baijiu. At present, the Chi flavor type Baijiu has been mechanized for its production, which greatly improves the production efficiency, reduces the labor intensity and stabilizes the product quality.

The Chi flavor type Baijiu is mainly produced in Guangdong Province, China. Its typical representative is the Shiwan Yubingshao Baijiu of Guangdong Shiwan Winery Group Co., Ltd.

52. Brewing of Te Flavor Type Baijiu

The Te flavor type Baijiu is a kind of Daqu Baijiu, which is crystal clear, with elegant flavor, pure taste and soft body. Its aroma is light and rich, elegant and comfortable, and with a mellow, sweet, round and innocent mouthfeel. The overall style of the Te flavor type Baijiu is summarized as having 'all three types but not really one of them', that is, with the characteristics of the Sauce, Strong and Light flavor types of Baijiu, but also having its own distinguishable characteristics.

The national standard defines the Te flavor type Baijiu as a kind of spirit with rice as its main raw material, a Daqu made from flour, wheat bran and distillers' grains as the saccharifying and fermenting starter, which is fermented by solid-state fermentation in a stone cellar of red tufted stripes, solid-state distillation, aging and blending, without directly or indirectly adding edible alcohol or non-self-fermentation substances.

The Te flavor type Baijiu is brewed by the solid-state technology of 'HZ-XCA' with the characteristics of 'whole grain rice as raw material; flour, bran and distillers' grains to prepare Daqu; and red stripe stone cellar'. 'Flour, bran and distillers' grains to prepare Daqu' refers to the fact that the Daqu is a saccharifying and fermenting starter for the Te flavor type Baijiu, and is made by mixing flour (35–40%), wheat bran (40–50%) and distillers' grains (20–15%) in a particular proportion, which is unique in the production of Baijiu. 'Red stripe stone cellar' refers to the fact that the fermentation cellar of the Te flavor type Baijiu is carried out in the red padded stone and cement joints, and mud is only used at the bottom of the cellar and in the closure of the cellar. The loose texture, large number of pores and strong water absorption of the red stripe stone are beneficial to the growth and reproduction of the brewing microorganisms, which is one of the reasons for the formation of the style of the Te flavor type Baijiu.

The Te flavor type Baijiu is a special product of Jiangxi Province in China. Typical representatives are Si'te Baijiu of Jiangxi Si'te Liquor Co., Ltd., Jiangxi Tequ Baijiu of Jiangxi Winery Co., Ltd., Lidu Baijiu of Jiangxi Lidu Liquor Co., Ltd., and Fuyun Tequ Baijiu of Jiangxi Fuyun Liquor Co., Ltd.

53. Brewing of Laobaigan Flavor Type Baijiu

'Laobaigan' means long history (lao), clear and transparent color (bai) and high alcohol content (gan). The Laobaigan flavor type Baijiu was officially recognized in 2007. It has the characteristics of pure, elegant and mellow flavor, and a rich and soft Baijiu body.

The national standard defines the Laobaigan flavor type Baijiu as a kind of spirit, which is made from grains by traditional solid-state fermentation, distillation, aging and blending, without adding edible alcohol or non-fermentation-derived substances, and has the composite flavor of ethyl acetate and ethyl lactate.

Laobaigan Baijiu is produced using sorghum as the starting material and middle-temperature Daqu as a saccharifying and fermenting starter. It is produced by the 'Lao Wu Zeng' technique (LWZ, which is a traditional process in the production of many flavor types of Baijiu in China. The marrow of this method is to steam and blend fermented grains five times with new grain materials. Under normal conditions, there are four layers of fermented grains in the cellar and five cycles of grain steaming and blending with new grain materials). Besides, the fermentation process is carried out in underground jars as shown in Figure 4.4, and the fermentation time is relatively short (12–14 days), while the yield of Baijiu is high (up to 50%). In addition, the storage aging period of Laobaigan Baijiu is short, generally 3–6 months. At present, the brewing technology has been mechanized for production.

Figure 4.4. Fermentation in Digang during the brewing process of the Laobaigan flavor type Baijiu.

The Laobaigan flavor type Baijiu is mainly produced in north and northeast China, and its typical representative is Laobaigan Baijiu and SBJF Baijiu from Hengshui, Hebei Province.

54. Brewing of Sesame Flavor Type Baijiu

The Sesame flavor type Baijiu is one of the two Baijiu flavor types created after the founding of the People's Republic of China. It is generally colorless to yellowish, crystal clear, with elegant and prominent sesame flavor, mellow and delicate, harmonious flavor and long-lasting aftertaste. The flavor is among the Strong, Light and Sauce flavor types.

According to the national standard, the Sesame flavor type Baijiu is made from sorghum, wheat (bran) and other grain materials by traditional solid-state fermentation, distillation, aging and blending, without adding edible alcohol or non-fermentation substances, and has a sesame flavor.

The Sesame flavor type Baijiu is brewed by the solid-state technology of 'HZ-XCA' or 'QZ-XCA', which has the characteristics of 'combination of Duaqu and Fuqu, and three high and one long'. 'Combination of Duaqu and Fuqu' refers to the use of mixed medium- and high-temperature Daqu and Fuqu as the saccharifying and fermenting starter. 'Three high and one long' refers to high nitrogen ingredients, high-temperature accumulation, high-temperature fermentation and a long aging time. In the brewing of the Sesame flavor type Baijiu, sorghum is the major ingredient along with wheat and bran. Because bran is rich in protein, the addition of it can improve the nitrogen–carbon ratio of the fermentation raw materials. 'High-temperature accumulation' refers to the accumulation of fermented grains before fermentation in the cellar, and the temperature can reach as high as 40~45°C. 'High-temperature fermentation' means that the fermentation temperature of grains is higher than that for the Light and Strong flavor types of Baijiu in the cellars. Generally, the fermentation temperature can go above 40°C and last for more than three days. The Sesame flavor type Baijiu is generally stored for 2–3 years so as to obtain stable sesame flavor characteristics.

The Sesame flavor type Baijiu is produced in Shandong, Jiangsu, Heilongjiang and Jilin Provinces and a few other places. Typical representatives are Guojing Baijiu of Shandong Guojing Holding Group Co., Ltd., Jingzhi Baijiu of Shandong Jingzhi Liquor Co., Ltd., Meilanchun Baijiu of Taizhou Meilanchun Liquor Factory Co., Ltd., Baotuquan Baijiu of Jinan Baotuquan liquor-making Co., Ltd., and Wuyue Duzun Baijiu of Taishan Liquor Group Co., Ltd.

55. Brewing of Fuyu Flavor Type Baijiu

The Fuyu flavor type Baijiu, also known as the Jiugui flavor type Baijiu, is one of the two major flavor types developed after the founding of the People's Republic of China. It has the characteristics of clear and transparent color, rich flavor, sweet and soft beginning taste, mellow and rich body, harmonious flavor and long aftertaste.

The Fuyu flavor type Baijiu is defined as a type of Baijiu with a unique style of pre-strong flavor, mid-light flavor and post-sauce flavor.

The Fuyu flavor type Baijiu is brewed using sorghum, rice (indica rice, polished round-grained rice, glutinous rice), wheat and maize as the four kinds of starting ingredients, and produced by high-temperature soaking and steaming of the grains separately, Xiaoqu saccharification, Daqu fermentation, low-temperature cellar entry, flavor enhancement in cellar mud, cave storage, careful blending procedures. The production of this flavor type of Baijiu is a solid-state brewing process by 'QZ-QS', which combines Xiaoqu and Daqu saccharifying and fermenting starters. Sorghum is the main grain material, accounting for 40% of the total grain ingredients. Xiaoqu is made from rice flour, and the main microorganism is pure rhizopus. Daqu is made from wheat at a temperature as high as 57–60°C, which belongs to a medium- to high-temperature Daqu.

The Fuyu flavor type Baijiu is mainly produced in the western Hunan Province, China. The typical representatives are Xiangquan Baijiu, Jiugui Baijiu and Neican Baijiu of Jiugui Liquor Co., Ltd., and they are geographical products of China.

56. Huangjiu-making in Jiangsu and Zhejiang Provinces

Jiangsu and Zhejiang Provinces are the major areas of Huangjiu production and consumption in China, accounting for more than 70% of the total.

The Huangjiu in Jiangsu and Zhejiang Provinces is made from rice (including japonica rice, glutinous rice and indica rice), by yeast fermentation with Maiqu as the saccharifying and fermenting agent after compression, precipitation, filtration, sterilization, filling in jars and aging. Huangjiu manufacturing can be carried out by traditional and mechanized techniques. With the advances in alcohol-making techniques, Huangjiu-making becomes more environmentally friendly, with high quality and efficiency, and using artificial intelligence. The fermentation process of traditional Huangjiu production and the

Figure 4.5. The fermentation process of traditional Huangjiu production.

Figure 4.6. The pre-fermentation process of mechanized Huangjiu production.

Figure 4.7. The post-fermentation process of mechanized Huangjiu production.

pre-fermentation and post-fermentation equipment for mechanized Huangjiu production are, respectively, shown in Figures 4.5–4.7.

Most of the Huangjiu produced in Zhejiang Province belongs to the traditional type, while in Jiangsu, it is the light (refreshing) type. The flavor of the light type comes from the yeasts and enzymatic preparations for the saccharifying, and less from Maiqu, and the taste is refreshing.

There are many Huangjiu brands in Jiangsu and Zhejiang Provinces and each has its unique manufacturing technique. According to the type of Qu used in Huangjiu-making, Huangjiu types are classified into Maiqu and Miqu.

(1) *Huangjiu fermented by Maiqu*

In the national standard *Huangjiu (GB/T 13662-2008)*, Huangjiu is classified into dry Huangjiu, semi-dry Huangjiu, semi-sweet Huangjiu and sweet Huangjiu according to the sugar content.

(1.1) *Dry Hangjiu*

Dry Huangjiu (sugar content ≤15 g/L) is made by the Tan-fan method, the Lin-fan method or the Wei-fan method depending on the different production areas. More water is used so that the concentration of the fermentation mixture is decreased, which along with the low fermentation temperature and the short interval of stirring and cooling accelerates the yeast fermentation of the raw materials, and ensures less residual starch, dextrin and sugar in the products and the dry taste. The typical brand for the Tan-fan method is the Shaoxing Yuanhong Huangjiu.

Huangjiu from Jiaxing is the typical kind of Huangjiu made by the Wei-fan method. The 'Wei-fan method' means that the raw materials are divided into several batches. The first batch is made into the mother yeast by the Lin-fan method and the other ingredients are added in several batches to ensure the continuous cultivation of yeast and continuous fermentation. This modern fermentation method created according to the law of microbial reproduction and fermentation is the same in principle as the 'nine-time addition' alcohol process used by Cao Cao in the Eastern Han Dynasty and the third-time, five-time and seven-time addition recorded *in Qi Min Yao Shu (Important Arts of the People's Welfare).*

(1.2) *Semi-dry Huangjiu*

Semi-dry Huangjiu (15 g/L < sugar content ≤40 g/L) is also called Jiafan Huangjiu because of the smaller amount of water and the relatively larger ratio of rice. Jiafan Huangjiu is classified into single and double Jiafan according to the added amount of rice. Jiafan Huangjiu is well made and of high quality, especially Shaoxing Jiafan Huangjiu; the amber liquid is glistening with a strong flavor, and a fresh and pure taste.

(1.3) *Semi-sweet Huangjiu*

The sugar content of a semi-sweet Huangjiu is between 40 g/L and 100 g/L, which results from the replacement of water with alcohol in the process. Similar to the 'mother–son' processing where water is replaced by soy sauce, the water is replaced by Yuanhong alcohol in the production of Shaoxing Shanniang Huangjiu. The input of alcohol instead of water at the beginning of fermentation may inhibit the growth of yeast to some degree and bring about incomplete fermentation. The high sugar content and other elements added with the aroma of the original Huangjiu bring a moderate alcohol content, sweet taste and special flavor to the semi-sweet Huangjiu.

(1.4) *Sweet Huangjiu*

Sweet Huangjiu (sugar content >100 g/L) is produced by the Lin-fan method. Saccharifying and fermenting agents are added into the cool cooked rice. Then, Baijiu of 40–50% alcohol concentration is added to inhibit the fermentation of yeast and keep a high sugar content. Sweet Huangjiu is always made in the summer because the initial alcohol content may protect against contamination from other undesirable microorganisms. Sweet Huangjiu products from different regions taste differently due to the different ingredients and manufacturing techniques. The representative sweet Huangjiu includes Xiangxue Huangjiu from Shaoxing of Zhejiang Province and Fenggang Huangjiu from Danyang of Jiangsu Province.

(2) *Huangjiu fermented by Miqu*

Miqu includes red Qu, black-coat red Qu and yellow-coat red Qu that make, respectively, red Qu Huangjiu, black-coat red Qu Huangjiu and yellow-coat red Qu Huangjiu. Huangjiu fermented by Miqu in Jiangsu and Zhejiang Provinces is mainly made into black-coat red Qu Huangjiu.

Made from indica rice, the black-coat red Qu containing monascus, aspergillus niger and yeast is a special saccharifying and fermenting agent from Huangjiu production in China. Black-coat red Qu Huangjiu originated in Wenzhou, Zhejiang Province, and was transferred to the southern Zhejiang regions such as Yiwu, Lishui, Quzhou and several areas in the nearby Fujian Province in the early 1970s.

The raw material of black-coat red Qu Huangjiu is indica rice, which is hard to steam thoroughly. So, the cooking procedure of 'steam twice and pour twice' is used. In recent years, with the development of indica rice steaming techniques, some distilleries smash the soaked indica rice for easy steaming, and loosen and spread it out for cooling.

Before being transferred into the jar, the black-coat red Qu is soaked to allow better release of the enzymes and other soluble substances, which accelerates the propagation of yeast. Soaking Qu is a vital step that affects the productivity and quality of the Huangjiu product.

The soaked Qu is mixed with the rice or ground rice, and fermented in jars for 10–15 days before further compression, sterilization and storage.

57. Huangjiu-making in Fujian Province

Fujian Huangjiu, also known as Fujian Aged Huangjiu or red Qu Huangjiu, is made by a traditional unique procedure using first-class glutinous rice; it uses medicinal Baiqu made from red Qu from the Gutian region and a Chinese medicinal herb formula as the saccharifying and fermenting agents, and requires a long-time fermentation at a low temperature in natural climate to enhance the flavor of Huangjiu, followed by a storage in jars for 3–5 years after compression and sterilization. The processing technique contributes to its unique sensory characteristics and quality. Figure 4.8 shows two red qu Huangjiu products from Fujian Province.

The origin of red Qu Huangjiu can hardly be confirmed, but the earliest written record has described it to certain level. In the 12th writing of volume 26 of *Primary Learnings* by Jianji Xu in the Tang Dynasty, it was recorded that 'Can Wang wrote in the *Seven Interpretations* that I travelled west to Liang and rested in Sucan. Red Qu Huangjiu in Guazhou tasted soft rich, smooth and refreshing'. This record from 2000 years ago shows that red Qu Huangjiu had been popular in the Eastern Han Dynasty.

Huangjiu can be classified into three groups of spicy Jiupei (dry), sweet Jiupei (sweet) and semi-spicy Jiupei (between the dry and sweet flavor) because of the different ingredients. Fujian Aged Huangjiu, a semi-sweet red Qu Huangjiu, is famous for its bright brown red color,

Figure 4.8. Different red qu Huangjiu products from Fujian Province.

strong and rich flavor, and pure and refreshing taste. The production process generally includes the following:

(a) Soaking the rice: The glutinous rice is soaked in the water for 8–12 hours in spring and winter, while 5–6 hours in summer.
(b) Washing the rice: Washing the soaked rice till the water is not turbid and draining off the water.
(c) Steaming or cooking the rice: Steaming or cooking the rice till it is well-cooked but not overcooked.
(d) Rapid cooling: Scattering the rice for cooling. The temperature is decided by that needed for mixing with Qu in the jar.
(e) Mixing with Qu in the jar: The jar is sterilized by steaming. After cooling down the temperature using water, red Qu is placed in the jar and soaked for 7–8 hours. The rice and Baiqu powders are added and mixed well in a jar, followed by placing the red Qu on the surface, and finally the jar is sealed with paper. In general, the temperature is kept in the range from 24 to 26°C for the mixture in the jar.

(f) Saccharification and fermentation: Generally, after 24 hours in the jar, the temperature of fermentation starts to increase. After 72 hours, the temperature rises and it is necessary to add water into the jar to keep the temperature from 35–36°C. Subsequently, the temperature goes down gradually and is close to an ambient temperature 7–8 days later. This is the pre-fermentation (chief) period.

(g) Stirring: Stirring is performed when necessary based on the observation of the fermented liquids. A stirring is needed when the surface of the fermentation mixture is thin and soft, or when an acrid aroma comes out from the mixture, or the liquids taste a little spicy or sweet, or the surface of the liquid sinks and cracks. The liquids are mature after being fermented for 90–120 days.

(h) Filtration using a framed filter: The mature fermentation liquids are filtered through a framed filter with pressing to obtain fresh Huangjiu.

(i) Clarifying, sterilization and filling: The filtered liquid is settled until clear, and subjected to sterilization and bottle filling.

58. Huangjiu Brewing in Daizhou

Huangjiu produced in Daizhou (Dai County) uses the typical manufacturing technique in the north of China. Huangjiu brewing techniques had originated and matured in and around Yangmingbao, Dai County, in the period of the Ming and Qing Dynasties.

Huangjiu in Daizhou, as shown in Figure 4.9, is made using millets, sorghum, mung beans, wine beans, red dates and Chinese

Figure 4.9. Daizhou (Dai County) Huangjiu from Shanxi Province.

wolfberries as the ingredient materials, and Daqu as the sacchariferous and fermentative agent. The major production procedures are as follows:

(a) Ingredients selection: First-class local millets are peeled into yellow millets, and mixed with mung beans, crystal sugar, quality red dates and liquor beans.
(b) Preparation of the Qu: The ingredients are soaked and steamed. After the resultant mixture is cooled down to certain degree, the starter is mixed in. The Qu is made through fermentation under certain temperature and moisture.
(c) Fermentation: The yellow millets are soaked and boiled in a pot with appropriate amount of water. After cooling to ambient temperature, the cooked millets are scattered and mixed with the Jiuqu starter, and transferred into a jar. The jar is sealed for fermentation. The fermentation temperature is examined and water is added into the jar at the correct moments to ensure that the fermentation progresses. After fermentation, the mixture is filtered using sandbags to collect the original Huangjiu, which is sealed and stored for further maturity.
(d) Final Huangjiu production: The original Huangjiu liquid is mixed with caramel syrup (made by heating crystal sugar), dates and liquor beans at a certain ratio. The mixture is boiled with water, followed by a blending procedure. The products are bottled for commercial markets after aging for a period of time. The more time for aging, the purer the liquids are, with a deep yellow and transparent appearance with no impurities, and a sweet and mild taste.

59. Aged Huangjiu from the Jimo Region of Shandong Province

Jimo aged Huangjiu, as Figure 4.10 shows, from Shandong is made from millets, Chenfu Maiqu and Laoshan mountain spring water following the ancient six critical control points in the liquor-making procedure: 'enough millets, timely addition of Qu, sweet spring

Figure 4.10. Jimo aged Huangjiu from Shandong Province.

water, quality pottery, clean steaming and full aging.' Jimo aged Huangjiu is obtained by natural fermentation followed by pressing filtration. It has a bright reddish color, strong aroma, unique flavor, mild feeling and pure taste.

Jimo aged Huangjiu is made from peeled millets. Compared with the Huangjiu made from rice, the production methods are different. For instance, the millets are boiled instead of steamed, and the addition of Qu and yeast makes the process similar to Baijiu fermented by Fuqu.

The main steps for making Shandong Jimo aged Huangjiu are as follows:

Millets → wash millets → boil millets (boiled water) → cooling → soaking → boiling till tender → cool → mix with Qu for saccharification → adding yeast → fermentation in jars → pressing → clarifying → sterilization → obtaining products.

Chapter 5
Flavors of Baijiu Products

60. Flavor Types of Baijiu Products

Different raw materials, saccharification and fermentation agents, fermentation equipment, fermentation and distillation processes, production and storage environments, and blending result in very diverse flavor types of Chinese Baijiu. Baijiu produced from different grain materials varies greatly in flavor style. It is generally accepted that 'sorghum produces fragrance, corn produces sweetness, rice produces purity, glutinous rice produces softness, and wheat produces spicy' flavor and mouthfeel for Baijiu products. In addition, the diverse microbial components in the large varieties of saccharification and fermentation agents have not only affected the flavor but also the alcohol yield of Baijiu.

Based on overall flavor characteristics, Chinese Baijiu can be classified into 12 flavor types, including Strong (Nong), Light (Mild), Sauce (Jiang), Rice (Mi), Feng, Mixed (Jian or Nongjiang), Dong, Chi, Te, Laobaigan, Sesame and Fuyu flavor types. Strong, Light, Sauce and Rice flavor types are the four primary flavors of Chinese Baijiu. The combination of Strong and Sauce flavor results in the Mixed flavor. The combinations of Strong and Light produces Feng flavor. Te and Fuyu flavors come from the combination of Strong, Light and Sauce flavors. Sauce, Rice and Strong-Sauce-Rice flavors generate Sesame, Chi and Dong flavors, respectively. The Laobaigan

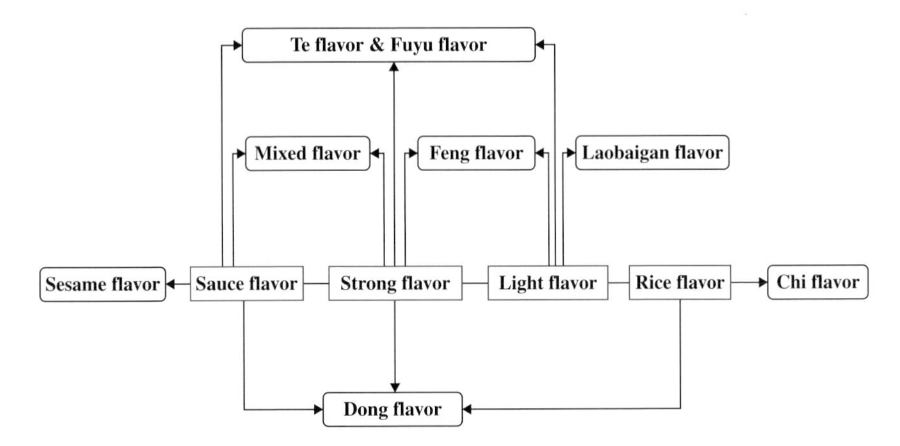

Figure 5.1. The relationships among different flavor types of Baijiu.

flavor type Baijiu is derived from the Light flavor type Baijiu. More details about the relationships among different flavor types of Baijiu are shown in Figure 5.1.

Although there are many descriptions of Baijiu flavor characteristics, the real aroma and taste profiles of Baijiu can only be realized after tasting.

61. Flavors of Strong Flavor Type Baijiu

The typical Strong flavor type Baijiu is rich in a harmonized cellar aroma and taste, characterized by ethyl hexanoate as the primary volatile component, a soft mouthfeel and a sweet taste, a sweet beginning and a long aftertaste.

There are two typical schools of aroma and taste profiles in the Strong flavor type Baijiu, including the Jianghuai School with an elegant strong flavor, and commonly produced in Jiangsu, Shandong, Henan and Anhui Provinces, and the Sichuan school with a strong flavor. The former school is typically represented by Yanghe Daqu Baijiu, Shuanggou Daqu Baijiu and Gujinggong Baijiu products, with a characteristic flavor from ethyl hexanoate and a pure taste, so that this school is called pure strong school. The latter school is represented by Luzhou Laojiao Baijiu, Wuliangye Baijiu and Jiannanchun

Figure 5.2. Five famous products of the Strong flavor type Baijiu.

Baijiu, with a characteristic of a well-balanced strong cellar aroma, and a well-developed flavor. Figure 5.2 shows five famous brands of the Strong flavor type Baijiu.

Reportedly, a total of 861 volatile compounds are identified in the Strong flavor type Baijiu, including 248 esters, 13 lactones, 122 alcohols, 65 acids, 53 aldehydes, 83 ketones, 35 acetals and ketals, 63 alkanes, 24 phenols, 16 ethers, 29 sulfur-containing compounds, 71 nitrogen-containing compounds, 35 furans and four other compounds. Ethyl acetate, ethyl butyrate, ethyl lactate, ethyl hexanoate, 1-propanol, isobutanol, isoamyl alcohol, phenylethanol, acetic acid, butyric acid, lactic acid, hexanoic acid, acetaldehyde, acetal, furfural, 2,6-dimethylpyrazine and tetramethylpyrazine are the most common organic compounds in the Strong flavor type Baijiu products.

Esters are the most abundant flavor components in the Strong flavor type Baijiu compared with other aroma compounds, in which the amount of ethyl hexanoate is often the highest besides ethanol and water. The odor threshold of ethyl hexanoate is 0.76 mg/L. This compound also has a sweet and refreshing taste. In brief, ethyl hexanoate determines the primary flavor feature of the Strong flavor type Baijiu. Ethyl hexanoate, together with the other three major esters including ethyl lactate, ethyl acetate and ethyl butyrate, contributes about 10–200 mg in each 100 mL of the Strong flavor type Baijiu. The ratio of the other esters to ethyl hexanoate is very important in determining the flavor of the Strong flavor type Baijiu, especially the ratios of ethyl lactate, ethyl acetate and ethyl butyrate to ethyl hexanoate, which partly determine the aroma and taste quality of the Strong flavor type Baijiu product.

Organic acids are important taste compounds in the Strong flavor type Baijiu, whose absolute contents are next to the esters at a level of about 140 mg/100 mL, about 25% of the total esters. Among all the organic acids, the levels of acetic acid, hexanoic acid, butyric acid and lactic acid are more than 10 mg/100 mL. The ratio of the total acids to esters needs to be in an appropriate range to ensure an excellent aroma and taste of Baijiu. In addition, the organic acids with higher boiling points are important for the duration of the aftertaste and after-flavor of the Baijiu.

Alcohols are also the important taste-contributing compounds in the Strong flavor type Baijiu with a total content about 103 mg/100 mL. The content of each alcohol compound may differ, with isoamyl alcohol as the most abundant one with a concentration range of 30–50 mg/100 mL. Sec.-butyl alcohol, isobutanol and *n*-butanol taste bitter, and a Baijiu sample may taste bitter when the levels of these compounds become greater. It is also noted that Baijiu products may have a bad taste when the concentration of isoamyl alcohol becomes higher. The concentrations of alcohols with a long carbon chain and polyhydric alcohols are generally low in Baijiu as they have low volatility with a sweet taste. These compounds contribute a more soft, sweet and full-bodied sensory property to the Baijiu products.

The contents of carbonyl compounds, acetals and ketals are low in the Strong flavor type Baijiu. Acetaldehyde and acetal are most abundant among this group of components with a level of more than 10 mg/100 mL followed by butanedione, 3-hydroxy-2-butanone and isovaleraldehyde, with concentrations from 4 to 9 mg/100 mL. The levels of propionaldehyde and isobutyraldehyde are less than the above compounds, which is 1–2 mg/100 mL. Butanedione and 3-hydroxy-2-butanone are volatile and have particular flavors. Their interactions with esters provide the Baijiu products with plentiful flavor. The greater the concentrations of butanedione and

3-hydroxy-2-butanone in certain concentration ranges, the better the flavor quality of a Strong flavor type Baijiu product.

62. Flavors of Light Flavor Type Baijiu

The characteristics of the Light flavor type Baijiu are pure, elegant and harmonious flavor with ethyl acetate as the primary contributor, with a slightly sweet entrance, a long after-flavor, a dry and refreshing mouthfeel and a slightly bitter taste. Figure 5.3 shows five famous brands of the Light flavor type Baijiu.

Reportedly, a total of 663 volatile compounds have been identified in the Light flavor type Baijiu, including 162 esters, 12 lactones, 93 alcohols, 45 acids, 49 aldehydes, 67 ketones, 32 acetals and ketals, 76 alkanes, 21 phenols, 13 ethers, 22 sulfur-containing compounds, 24 furans, 39 nitrogen-containing compounds and eight other components. Ethyl acetate, ethyl butyrate, ethyl lactate, ethyl valerate, diethyl succinate, 1-propanol, isobutanol, isoamyl alcohol, phenylethanol, acetic acid, butyric acid, lactic acid, butanedioic acid, acetaldehyde, acetal, furfural and tetramethylpyrazine are the most common organic compounds in the Light flavor type Baijiu. The concentrations of esters are the most abundant followed by that of alcohols, acids, aldehydes, ketones and heterocycles in the Light flavor type Baijiu. The total concentrations of aroma compounds in the

Figure 5.3. Five famous products of the Light flavor type Baijiu.

Light flavor type Baijiu are generally lower than that in the Strong flavor type Baijiu.

The amount of ethyl acetate is the highest among all the esters, with a concentration of more than 200 mg/100 mL. Ethyl lactate is the second highest content ester at a level above 100 mg/100 mL. The concentration of ethyl acetate and ethyl lactate and the ratio of these two compounds (generally 1: 0.6–0.8) are important to the overall flavor profiles of the Light flavor type Baijiu. In addition, diethyl succinate is another very important aroma compound among the esters, and it can interact with β-phenylethanol and result in a sweet flavor of the Baijiu products.

Organic acids have an important influence on the taste and mouthfeel of the Light flavor type Baijiu. Acetic and lactic acids are the primary acids with the content level above 20 mg/100 mL. Furthermore, the ratio of acetic to lactic acid is about 1: 0.8. The concentrations of acetic and lactic acids account for more than 90 percent of the total acids, and other acids are in low levels. The amounts of the total acids are generally from 20 to 120 mg/mL. A too high or too low content of the total acids is not favored for an excellent flavor of the Baijiu.

Alcohols are also important flavor compounds in the Light flavor type Baijiu, with a level of about 67 mg/100 mL. The concentrations of isoamyl alcohol, 1-propanol and isobutanol are most abundant in terms of the alcohols. Alcohols have a relative high ratio of the total volatile compounds, which is a feature of the Light flavor type Baijiu. The flavor feature of the Light flavor type Baijiu is a slight sweetness at the beginning, and powerful astringency; additionally, it is refreshing. This characteristic flavor feature is largely associated with the alcohol compounds.

The amounts of carbonyl compounds, acetals and ketals in the Light flavor type Baijiu are not high, with acetaldehyde and acetal as the most abundant ones, accounting 90% of the total. Acetaldehyde and acetal possess a strong pungent smell, especially acetal with a refreshing taste, which give Baijiu a refreshing and bitter feeling when they interact with 1-propanol.

63. Flavors of Sauce Flavor Type Baijiu

The typical Sauce flavor type Baijiu is described as an elegant sauce flavor, empty cup aroma, lasting, mellow and sweet entrance, soft with obvious acidity, delicate taste and long aftertaste. Figure 5.4 shows five famous brands of the Sauce flavor type Baijiu.

As reported, a total of 623 volatile components are identified in the Sauce flavor type Baijiu, including 163 esters, 11 lactones, 66 alcohols, 47 acids, 34 aldehydes, 54 ketones, 23 acetals and ketals, 62 alkanes, 29 phenols, 8 ethers, 17 sulfur-containing compounds, 38 furans, 67 nitrogen-containing compounds and four other compounds. Ethyl acetate, ethyl butyrate, ethyl lactate, ethyl hexanoate, ethyl palmitate, 1-propanol, isobutanol, isoamyl alcohol, β-phenylethanol, acetic acid, butyric acid, lactic acid, hexanoic acid, acetaldehyde, acetal, furfural, tetramethylpyrazine and trimethylpyrazine are the most common organic compounds in the Sauce flavor type Baijiu.

Compared with other flavor types of Baijiu, the flavor components of the Sauce flavor type Baijiu can be characterized as 'three high, one low, and two more', which refers to high contents of acids, alcohols, aldehyde and ketone; low content of ester; and more abundance of high boiling point compounds and heterocyclic substances.

The total concentration of organic acids in the Sauce flavor type Baijiu is about 300 mg/100 mL, which is much higher than that of

Figure 5.4. Five famous products of the Sauce flavor type Baijiu.

the Strong or Light flavor types of Baijiu. Among all the organic acids, the levels of acetic acid and lactic acid are more than 100 mg/100 mL, which is greater than the other acids, and are the highest in all flavor types of Baijiu.

The level of alcohols is about 270 mg/100 mL, with propanol as the highest at a level about 140 mg/100 mL. The high concentration of alcohols may enhance the flavor of other volatile compounds.

The concentration of esters in the Sauce flavor type Baijiu is less than that in other flavor types of Baijiu, and is about 40–50 mg/100 mL. The most abundant ones are ethyl acetate and ethyl lactate.

The total levels of aldehydes and ketones in the Sauce flavor type Baijiu are the highest among all flavor types of Baijiu, especially for furfural with a concentration of 29.4 mg/100 mL.

There are many compounds with high boiling points in the Sauce flavor type Baijiu. Their levels are the highest in all flavor types of Baijiu. These compounds include high boiling point organic acids, alcohols, esters, aromatic acids and amino acids. They can soften the taste of the Sauce flavor and keep the aroma of an empty cup for a long time, and thus, are the key flavor compounds in the Sauce flavor type Baijiu.

The Sauce flavor type Baijiu has a high level of nitrogen-containing heterocyclic compounds (6.43 mg/100 mL), which is the highest among all flavor types of Baijiu. These compounds have diverse chemical structures. The pyrazine compounds are the most abundant in the group. The concentration of tetramethylpyrazine can be as high as 5.30 mg/100 mL.

64. Flavors of Rice Flavor Type Baijiu

The typical Rice flavor type Baijiu products have a mixed ethyl acetate- and β-phenylethanol-based elegant and composite flavor. The Baijiu tastes sweet and mellow with a pleasant and refreshing mouthfeel. The Baijiu samples may taste a little bitter and the flavor does not last long. Figure 5.5 shows three famous brands of the Rice flavor type Baijiu.

Figure 5.5. Three famous products of the Rice flavor type Baijiu.

The Rice flavor type Baijiu is produced through a relatively simple brewing process and shorter fermentation period using a semi-solid fermentation technique, and results in less flavor components with a mild flavor. A total of 109 volatile compounds have been identified in the Rice flavor type Baijiu, including 34 esters, 21 alcohols, 19 acids, 10 aldehydes, 9 ketones, 6 phenols and 10 other compounds. Ethyl acetate, ethyl lactate, ethyl palmitate, ethyl oleate, 1-propanol, isoamyl alcohol, isobutanol, acetic acid, lactic acid, heptanic acid, acetaldehyde, acetal and furfural are the common organic compounds in the Rice flavor type Baijiu.

The amount of ethyl lactate is up to 100 mg/100 mL, which has exceeded the concentration of ethyl acetate (about 25 mg/100 mL); therefore, the Baijiu has a little bitter taste. This is also a key factor of the elegant flavor.

The total alcohol contents (about 170 mg/100 mL) in the Rice flavor type Baijiu are more than the total ester contents (about 134 mg/100 mL), and result in the mellow flavor and the bitter taste of the Baijiu. The content of β-phenylethanol (about 3.3 mg/100 mL) in the Rice flavor type Baijiu exceeds that in the Light or Strong flavor types of Baijiu. The proportion of β-phenylethanol has been greatly increased due to the lower concentrations of the total aroma compounds in the Rice flavor type Baijiu. Additionally, the low threshold of β-phenylethanol makes the ethyl acetate- and β-phenylethanol-based elegant and composite flavor a typical feature of the Rice flavor type Baijiu.

The thickness taste of the Rice flavor type Baijiu is not stronger than that of the Strong or Light flavor types of Baijiu because of having fewer organic acids. The flavor of the Rice flavor Baijiu cannot last long due to the same reason. Lactic acid is the primary acid with a content of about 100 mg/100 mL in the Rice flavor type Baijiu, followed by that of acetic acid at a level of about 22 mg/100 mL. The total content of lactic and acetic acids accounts for more than 90 percent of the total acids.

65. Flavors of Feng Flavor Type Baijiu

A typical Feng flavor type Baijiu is colorless, clear and transparent. The Feng flavor type Baijiu has a unique mellow flavor. The multiple flavors, mainly from ethyl acetate and a certain amount of ethyl caproate and other esters, are faint. The taste is full-bodied and powerful, with a pleasant and refreshing feeling after drinking. Figure 5.6 shows two famous brands of the Feng flavor type Baijiu.

A total of 109 volatile components have been identified in the Feng flavor type Baijiu, including 36 esters, 19 alcohols, 18 acids, 6 aldehydes, 12 phenols, 8 furans and 10 other compounds. Ethyl acetate, ethyl butyrate, ethyl lactate, ethyl hexanoate, ethyl palmitate,

Figure 5.6. Two famous products of the Feng flavor type Baijiu.

1-propanol, isobutanol, isoamyl alcohol, beta-phenylethanol, acetic acid, butyric acid, lactic acid, hexanoic acid, acetaldehyde, acetal, furfural, tetramethylpyrazine and p-cresol are the common flavor compounds in the Feng flavor type Baijiu. The flavor profile of the Feng flavor type Baijiu are somewhere between the Strong and Light flavor types of Baijiu products. The concentrations of total acids and esters are, respectively, 77 mg/100 mL and 160–280 mg/100 mL, which are lower than those in the Strong and Light flavor types of Baijiu, while the total alcohol (about 130 mg/100 mL) is more than that in the two flavor types of Baijiu.

The amount of ethyl acetate is the highest in the Feng flavor type Baijiu, generally ranging from 80 to 150 mg/100 mL, which is lower than that in the Strong and Light flavor types of Baijiu. The concentration of ethyl hexanoate is higher than that in the Light flavor type Baijiu, while obviously lower than that in the Strong flavor type Baijiu. Alcohols are the abundant compounds in the Feng flavor type Baijiu, and the level of isoamyl alcohol is the highest at about 52 mg/100 mL, followed by that of *n*-butanol, *n*-propanol and isobutanol, whose concentrations are all at about 21 mg/100 mL. High amounts of alcohols contribute with a mellow flavor, which is a feature of the Feng flavor type Baijiu. However, the high amounts of alcohols give Baijiu a powerful taste. There is no significant difference between the total concentrations of aldehyde and ketone compounds (about 79 mg/100 mL) and total acids. The content of acetaldehyde is about 36 mg/100 mL.

66. Flavors of Mixed Flavor Type Baijiu

The so-called ‘Mixed flavor-type Baijiu’ refers to a flavor that has features of both the Strong and Sauce flavor types of Baijiu, and the two flavors combine harmoniously. Figure 5.7 shows three famous brands of the Mixed flavor type Baijiu.

There are two flavor groups in the Mixed flavor type Baijiu. One group is represented by Baiyunbian Baijiu, with a sauce-based flavor and a faint aroma of ethyl hexanoate after drinking with a long-lasting flavor. The other group is represented by Yuquan Baijiu, with a soy

Figure 5.7. Three famous products of the Mixed flavor type Baijiu.

sauce-like aroma and a faint aroma of ethyl hexanoate, a well-combined Sauce and Strong flavor, a symbolic flavor of ethyl hexanoate at the beginning, a soft and sweet mouthfeel and an aftertaste with sauce aroma.

A total of 171 volatiles have been identified in the Mixed flavor type Baijiu, including 55 esters, 23 alcohols, 15 acids, 10 aldehydes, 4 ketones, 6 acetals and ketals, 9 phenols, 3 sulphur-containing compounds, 9 furans and 37 nitrogen-containing compounds. The concentrations of many flavor compounds in the Mixed flavor type Baijiu, such as ethyl hexanoate, hexanoic acid, furfural, beta-phenylethanol, ethyl propionate, ethyl isobutyrate, 2,3-butanediol, isobutanol, isovaleric acid and pyrazines, are in the same ranges as their levels in the Strong and Sauce flavor types of Baijiu, explaining well why the Mixed flavor type Baijiu displays the typical features of the Strong and Sauce flavors. The content of ethyl hexanoate, an important aroma compound in the Mixed flavor type Baijiu, generally ranges from 60 to 120 mg/100 mL.

There is a clear difference between the two flavor groups of Baijiu. The concentration of ethyl hexanoate in the Yuquan Baijiu group is generally more than twofold that in the Baiyunbian Baijiu group. The amount of hexanoic acid is more abundant than acetic acid in the Yuquan Baijiu group, different from the Baiyunbian Baijiu group. In addition, the concentration of furfural in the Yuquan Baijiu group is generally greater than that in the Baiyunbian Baijiu by nearly 30%, and is almost 10 times greater than that in the Strong flavor type Baijiu, similar to that in the Sauce flavor type Baijiu. The content of beta-phenylethanol in Yuquan Baijiu is 23% higher than that in Baiyunbian Baijiu, also close to that of the Sauce flavor type Baijiu. The amount of diethyl succinate in Yuquan Baijiu is many times higher than that in Baiyunbian Baijiu.

67. Flavors of Dong Flavor Type Baijiu

The Dong flavor type Baijiu, also called the Herblike flavor type Baijiu, is represented by Dong Baijiu with rich flavors from esters with a medicinal herb incense aroma, and a composite flavor of butyric acid and ethyl butyrate. The Baijiu products taste sweet and mellow with a sour beginning note, and a long aftertaste. The flavor of esters, the mellowness and the medicinal herb are important for the Dong flavor type Baijiu, among which the herb incense is a primary characteristic. Figure 5.8 shows a typical representative of the Dong flavor type Baijiu.

A total of 138 volatiles have been identified in the Dong flavor type Baijiu, including 28 esters, 30 alcohols, 15 acids, 9 aldehydes, 9 ketones, 29 alkanes, 6 phenols and 12 other compounds.

Figure 5.8. A typical representative of the Dong flavor type Baijiu.

Ethyl acetate, ethyl butyrate, ethyl lactate, ethyl hexanoate, 1-propanol, isobutanol, isoamyl alcohol, acetic acid, butyric acid, pentanoic acid, hexanoic acid, lactic acid, acetaldehyde, acetal and butanedione are the common flavor compounds in the Dong flavor type Baijiu.

Ethyl acetate, ethyl butyrate, ethyl hexanoate and ethyl lactate are the four most important esters to provide the composite ester flavor, which differentiates the Dong flavor type Baijiu from the Strong and Light flavor types of Baijiu products; both have predominant esters for their flavor types. Moreover, the concentration of ethyl butyrate in the Dong flavor type Baijiu is 3–4 times greater than that in other flavor types of Baijiu, which is closely related to their elegant ester flavor and the full-bodied taste. The content of ethyl lactate is relatively low, at a level about 1/3–1/2 of that in other flavor types of Baijiu, and contributes to the dry and refreshing features of Dong Baijiu.

The total concentrations of alcohols in the Dong flavor type Baijiu is greater than that of the esters, and the ratio of alcohols to esters is greater than 1. This is similar to the Rice flavor type Baijiu that alcohols play important roles in the Dong flavor of Baijiu, another feature of the Dong flavor type Baijiu. In addition, 1-propanol (147 mg/100 mL), sec.-butanol (133 mg/100 mL) and isoamyl alcohol (93 mg/100 mL) are the major alcohol compounds. 1-propanol and sec-butanol contribute to the pleasant smell and elegant flavor, and can be combined with ester flavor, thus highlighting the unique elegant flavor characteristic of the Dong flavor type Baijiu.

The total concentration of organic acids is greater than the total ester content for the Dong flavor type Baijiu. Acetic acid (132.1 mg/100 mL), butyric acid (46.2 mg/100 mL), hexanoic acid (31.1 mg/100 mL) and lactic acid (49 mg/100 mL) are the common volatile acids in the Dong flavor type Baijiu. Compared with other flavor types of Baijiu, the content of butyric acid is 10 times higher in the Dong flavor type Baijiu, along with a higher concentration of ethyl butyrate, resulting in the typical flavor of butyric acid and ethyl butyrate, a flavor characteristic of the Dong flavor type Baijiu. In addition, the total esters in the Dong flavor type Baijiu are less than the total acids, which differentiates Dong Baijiu from other flavor types of Baijiu. The high concentrations of acids play an important role in the refreshing aftertaste of the Dong flavor type Baijiu.

The concentration ratios of the flavor compounds in the Dong flavor type Baijiu can be summarized as 'three high and one low'. The 'three high' is the high contents of ethyl butyrate, non-ethanol alcohols (especially *n*-propanol and sec-butanol) and total acids (especially butyric acid). The 'one low' reflects the low concentration of ethyl lactate.

68. Flavors of Chi Flavor Type Baijiu

The Chi flavor type Baijiu is represented by Yubingshao Baijiu produced in Guangdong Province with elegant flavors of ethyl acetate and β-phenylethanol, along with an obvious fat-oxidized bacon aroma. This type of Baijiu tastes soft and supple with a long aftertaste. It is a little bitter upon drinking without a mouthful left, and its aftertaste is refreshing. Figure 5.9 shows the typical representative of the Chi flavor type Baijiu.

The so-called 'Chi (its Chinese character stands for fermented soya beans) flavor' does not mean the flavors from fermented soya beans, but it is a special composite flavor from the soaking fatty meat obtained during the post-ripening process for producing the Rice

Yubingshao Baijiu

Figure 5.9. The typical representative of the Chi flavor type Baijiu.

flavor type Baijiu (also known as Zhai jiu). It is the unique flavor characteristic of Yubingshao Baijiu.

A total of 122 flavor volatiles have been identified in the Chi flavor type Baijiu, including 30 esters, 27 alcohols, 27 acids, 9 aldehydes, 3 ketones, 16 phenols and 10 other compounds. Ethyl acetate, ethyl butyrate, ethyl lactate, ethyl hexanoate, 1-propanol, isobutanol, isoamyl alcohol, acetic acid, butyric acid, pentanoic acid, hexanoic acid, lactic acid, acetaldehyde, acetal and butanedione are the common volatile compounds in the Chi flavor type Baijiu.

Compared with the Rice flavor type Baijiu, the Chi flavor type Baijiu has its characteristic flavor compounds. The relative content of β-phenethyl alcohol is high, ranging from 2.0 to 12.7 mg/100 mL with an average level of 6.6 mg/100 mL. This level is the highest in all flavor types of Baijiu, and is nearly twice the concentration of that in the Rice flavor type Baijiu. This is a characteristic of Chi Baijiu flavor components. In addition, the process of soaking fatty meat results in a higher concentration of carbonyl compounds and dibasic acid esters with high boiling points, mainly including diethyl heptanedioate (0.578–0.736 mg/L), diethyl azelate (1.61–1.70 mg/L) and diethyl octanediate (1.12–1.94 mg/L).

The flavor characteristic of the Chi flavor type Baijiu is similar to that of the Rice flavor type Baijiu, while also having its own unique properties. In teams of the aroma, it highlights the flavor of mellow aroma with an elegant ethyl acetate- and β-phenylethanol-based composite flavor because of the relatively higher alcohol contents and low ester contents. In addition, it also has a faint flavor of cooked rice. These are some similarity between the two types of Baijiu. The difference is the flavor of β-phenylethanol in the Chi flavor type Baijiu with an obvious stored meat flavor after fat oxidation, which is highlighted more than in the Rice flavor type Baijiu. Compared with the Rice flavor type Baijiu, the Chi flavor type Baijiu tastes softer, longer-lasting and less bitter due to the higher concentrations of high boiling point substances.

69. Flavors of Te Flavor Type Baijiu

The typical flavor characteristic of the Te flavor type Baijiu is ethyl acetate- and ethyl hexanoate-based composite ester flavor,

Figure 5.10. The typical representative of the Te flavor type Baijiu.

accompanied by light burnt aroma. It has obvious an ethyl heptanoate odor with a soft taste and long-lasting feeling after drinking; additionally, it has a distinct sweet taste. Figure 5.10 shows the typical representative of the Te flavor type Baijiu.

A total of 133 volatiles have been identified in the Te flavor type Baijiu, including 44 esters, 21 alcohols, 25 acids, 16 aldehydes and ketones, 6 acetals, 18 nitrogen-containing compounds, 2 sulfur-containing compounds and 1 other compound. Ethyl acetate, ethyl butyrate, ethyl lactate, ethyl hexanoate, ethyl heptanoate, ethyl palmitate, ethyl oleate, 1-propanol, isobutanol, isoamyl alcohol, β-phenylethanol, acetic acid, butyric acid, pentanoic acid, hexanoic acid, acetaldehyde, acetal, furfural, tetramethylpyrazine and trimethylpyrazine are the common volatile compounds in the Te flavor type Baijiu.

Esters are the most abundant flavor components in the Te flavor type Baijiu with about 352 mg/100 mL. Ethyl acetate (135 mg/100 mL), ethyl lactate (112 mg/100 mL) and ethyl hexanoate (32 mg/100 mL) are the major esters. In addition, ethyl hexanoate plays a critical role in the ester aroma of the Te flavor type Baijiu due to its low threshold. Ethyl hexanoate is also important for the flavor of the Strong flavor type Baijiu. In addition, the Te flavor type Baijiu is rich in fatty acid ethyl esters with odd carbon molecules. The concentration of total fatty acid ethyl esters in the Te flavor type Baijiu is the highest in all flavor types of Baijiu, which is another important feature of the Te flavor type Baijiu.

The alcohol contents are also higher among all the aroma compounds in the Te flavor type Baijiu, and are similar to the contents of esters. The concentration of 1-propanol is the highest ranging from 59 to 307 mg/100 mL. The concentration of 1-propanol in the Te flavor Baijiu is the highest among all flavor types of famous Baijiu products in China, and is 4–5 times higher than that in the other famous and quality types of Baijiu, and is another important character of the Te flavor type Baijiu.

The acid concentration of the Te flavor Baijiu ranks the third with a total value of 130 mg/100 mL. The level of acetic acid is the highest at 82 mg/100 mL, followed by the hexanoic acid, valeric acid, butyric acid and propionic acid. Moreover, the concentrations of longer chain fatty acids and the corresponding esters in the Te flavor type Baijiu are higher than in the other flavor types of Baijiu, often with palmitic acid and ethyl palmitate as the primary components at 2.46 mg/100 mL and 7.09 mg/100 mL, respectively. These substances play an important role in the soft taste and lasting flavor of the Te flavor type Baijiu.

The content of acetaldehyde is 17 mg/100 mL, which is the highest among the carbonyl compounds. The level of acetal is 24 mg/100 mL, which is the highest among all the acetals. The total concentration of nitrogen-containing heterocyclic compounds is about 0.18 mg/100 mL, among which pyrazine has the highest contents with the concentration of tetramethylpyrazine at about 0.06 mg/100 mL.

70. Flavors of Laobaigan Flavor Type Baijiu

The Laobaigan flavor type Baijiu is represented by Hengshui Laobaigan Baijiu produced in Hebei Province, with elegant mellow flavor and the composite aroma of ethyl lactate and ethyl acetate. The flavor of the final product is well-balanced, sweet and with a long aftertaste, which has a unique style of its own. Figure 5.11 shows the typical representative of the Laobaigan flavor type Baijiu.

Reportedly, a total of 544 volatiles have been identified in the Laobaigan flavor type Baijiu, including 167 esters, 78 alcohols,

Hengshui Laobaigan Baijiu

Figure 5.11. The typical representative of the Laobaigan flavor type Baijiu.

31 aldehydes, 54 ketones, 34 acids, 13 lactones, 36 nitrogen-containing compounds, 45 sulfur-containing compounds, 18 furans, 30 acetals, 10 phenols, 13 ethers, 20 hydrocarbons and 5 anhydrides. Ethyl acetate, ethyl lactate, ethyl hexanoate, 1-propanol, isobutanol, isoamyl alcohol, acetic acid, butyric acid, isovaleric acid and lactic acid are the common volatile compounds in the Laobaigan flavor type Baijiu.

Ethyl acetate and ethyl lactate are the major ester components with a ratio of ethyl acetate to ethyl lactate at 1: (1.5–2.0). In addition, ethyl hexanoate and ethyl butyrate, as well as longer chain fatty acid esters such as ethyl palmitate, are the other common esters. Acetic, lactic, valeric and hexanoic acids are the most abundant organic acids. 1-propanol, isobutanol and isoamyl alcohol are the major alcohols, and their contents are higher than that in Fen Baijiu, the representative Baijiu of the Light flavor type.

Among all flavor substances in Laobaigan Baijiu, 4-ethyl guaiacol, 2-phenylethyl acetate, butyric acid, 3-methyl butanol, beta-phenylethanol, 2-acetyl-5-methyl furan, ethyl phenylpropionate, gamma-nonolide, 3-methyl butyric acid, vanillin, ethyl acetate, 1,1-diethoxy-3-methyl butane and (2,2-diethoxyethyl) phenyl are major contributors for the overall flavor of the Laobaigan flavor type Baijiu.

71. Flavors of Sesame Flavor Type Baijiu

The Sesame flavor type Baijiu is a kind of innovative Baijiu with the characteristics of the Strong, Sauce and Light flavor types of Baijiu, but different from all of them. It has an elegant flavor and the aroma of ethyl acetate, the primary ester, with a prominent baking flavor. The entrance flavor mainly includes baking and pasting flavors, along with the aroma of 'roasted sesame'. This type of Baijiu tastes mellow and refreshing, which is like the taste of the Laobaigan flavor type of Baijiu, and has a slightly bitter aftertaste. Figure 5.12 shows five famous brands of the Sesame flavor type Baijiu.

A total of 299 volatile compounds have been reported in the Sesame flavor type Baijiu, including 72 esters, 46 alcohols, 36 acids, 14 aldehydes, 29 ketones, 5 acetals and ketals, 16 phenols, 23 sulfur-containing compounds, 47 nitrogen-containing compounds and 11 other compounds. Ethyl acetate, ethyl butyrate, ethyl lactate, ethyl hexanoate, *n*-propanol, isobutyl alcohol, isoamyl alcohol, acetic acid, butyric acid, valeric acid, hexanoic acid, lactic acid, acetaldehyde, acetal, trimethylpyrazine, tetramethylpyrazine, dimethyl trisulfide, ethyl 3-methylthiopropionate and 3-methylthiopropionaldehyde are the common flavor compounds in the Sesame flavor type Baijiu.

The Sesame flavor type Baijiu contains ethyl caproate (44.0 mg/100 mL) and hexanoic acid (26.1 mg/100 mL) at levels lower than that in the Strong and Mixed flavor types of Baijiu. These two

Figure 5.12. Five famous products of the Sesame flavor type Baijiu.

compounds contribute to the flavor characteristics of the Strong flavor type Baijiu. However, the levels of these two flavor compounds are slightly higher than those in the Sauce flavor type of Baijiu, and additionally significantly higher than those of the Light flavor type of Baijiu. The above features are accordant with its elegant aroma style.

The contents of ethyl acetate, diethyl succinate, *n*-propanol and isoamyl alcohol are 160 mg/100 mL, 0.40 mg/100 mL, 17.1 mg/100 mL and 33.2 mg/100 mL, respectively. These contents are similar to those in the Light flavor type Baijiu, and contribute some flavor features of the Light flavor type of Baijiu to the Sesame flavor type Baijiu.

Furfural, beta-phenylethanol and benzaldehyde are closely related to the Sauce flavor type Baijiu. The content of furfural (8.94 mg/100 mL) in the Sesame flavor type of Baijiu is lower than that in the Sauce and Mixed flavor types of Baijiu, but is greater than that in the Strong and Light flavor types of Baijiu. The content of beta-phenylethanol (1.36 mg/100 mL) is about the same as that in the Mixed flavor type Baijiu, but is obviously higher than that in the Strong flavor type Baijiu. The content of benzaldehyde (1.7 mg/100 mL) is higher than that of the Sauce flavor type Baijiu.

Overall, the flavor components in the Sesame flavor type Baijiu are similar to those in the Strong, Light and Sauce flavor types of Baijiu. However, the flavor of the Sesame flavor type Baijiu is quite different from the three types of Baijiu due to the different contents of some characteristic components and the relative ratios of the flavor compounds.

Sulphur-containing compounds, such as 3-methylthiopropionaldehyde, and dimethyl trisulfide provide an essential contribution to the overall flavor characteristics of the Sesame flavor type Baijiu.

72. Flavors of Fuyu Flavor Type Baijiu

The Fuyu flavor type Baijiu, represented by Jiugui Baijiu in Xiangxi, is characterized by elegant flavor, soft and sweet taste, mellow, delicate and pleasant aftertaste, rich aroma and refreshing body. The Fuyu

Figure 5.13. Three famous products of the Fuyu flavor type Baijiu.

flavor type Baijiu blends the flavors of the Sauce, Strong and Light flavor types of Baijiu ingeniously. It is also an innovative flavor type in the Chinese Baijiu industry. Figure 5.13 shows three famous brands of the Fuyu flavor type Baijiu.

More than 200 volatile compounds have been reported in the Fuyu flavor type Baijiu, in which ethyl acetate, ethyl butyrate, ethyl lactate, ethyl hexanoate, *n*-propanol, isobutyl alcohol, isoamyl alcohol, acetic acid, butyric acid, isovaleric acid, hexanoic acid, lactic acid, acetaldehyde, acetal and furfural are the common flavor compounds.

The total content of esters in the Fuyu flavor type Baijiu is higher than the other aroma compounds. In particular, the ethyl caproate and ethyl acetate contents are higher than the other esters (over 100–170 mg/100 mL). The contents of these two esters are comparable, but the content of ethyl acetate is slightly higher than that of ethyl caproate. This is a unique characteristic of the Fuyu flavor type of Baijiu, which is not found in other flavor types of Baijiu. Generally, the concentration of ethyl lactate ranges from 53 to 72 mg/100 mL, and that of ethyl butyrate is 16–29 mg/100 mL. The contents and ratios of the 'four esters' in the Fuyu flavor type Baijiu are very different from the Strong and Light flavor types of Baijiu, and Xiaoqu Baijiu made in Sichuan Province. It shows that the Fuyu flavor type Baijiu uses the Xiaoqu Baijiu-making technique instead of that for the Light flavor type Baijiu, and uses the technique of Daqu Baijiu but not the same as that for the Strong flavor type Baijiu. The creative processing technology results in its own unique flavor style.

The content of organic acids in the Fuyu flavor type Baijiu is also high at a total level of more than 200 mg/100 mL, which is much higher than that in the Strong and Light flavor types of Baijiu, and Xiaoqu Baijiu made in Sichuan Province, but is lower than that in the Sauce flavor type Baijiu. Caproic acid and acetic acid, lactic acid and butyric acid are the major acids and account for 70%, 19% and 7% of the total acids, respectively. Although the proportion of these four acids is similar to that of the Strong flavor type of Baijiu in the order of acetic acid > caproic acid > lactic acid > butyric acid, the contents of acetic acid and caproic acid are two times higher than those in the Strong flavor type of Baijiu. However, the types of organic acids in the Light flavor type Baijiu and Xiaoqu Baijiu made in Sichuan Province are single, which are obviously different from the rich types of organic acids in the Fuyu flavor type of Baijiu.

The contents of alcohols in the Fuyu flavor type Baijiu are moderate, ranging from 110 to 140 mg/100 mL, which are higher than those in the Strong and Light flavor types of Baijiu, but lower than those in the Xiaoqu Light flavor type Baijiu. The content of isoamyl alcohol is the highest among all the alcohols at about 40 mg/100 mL, followed by *n*-propanol at 25–50 mg/100 mL, which is lower than that in the Sauce, Dong and Te flavor types of Baijiu, but is higher than that of the products which are relatively close in brewing technology including the Strong, Light and Sichuan Xiaoqu flavor types of Baijiu.

Chapter 6
Famous Baijiu

73. Baofeng Baijiu

Baofeng Baijiu, as shown in Figure 6.1, is a typical representative of the Daqu Light flavor type Baijiu. It is produced in Baofeng County, Pingdingshan City, Henan Province, and is a product of Henan Baofeng Liquor Co., Ltd. The processing research of Baofeng Baijiu was started in the Shenzong period of the Northern Song Dynasty. According to the records of *Henan Tongzhi*, *Ruzhou Zhi* and *Baofeng County Zhi*, Hao Cheng, a Neo-Confucian scholar in the Northern Song Dynasty, once supervised Baijiu brewing and lectured in Baofeng, where there were more than 100 distilleries. In the Jiaqing period of the Qing Dynasty, according to the local record *Baofeng Zhi*, the tax on alcohol in Baofeng country reached 45,000 strings in 1224 A.D., ranking the first among all counties in China. At present, Baofeng Baijiu is the only Light flavor type Baijiu produced by above-scale enterprises in Jiangsu, Shandong, Henan and Anhui Provinces. The brewing process of Baofeng Baijiu follows the traditional 'Qingzheng Erciqing' (QZ-ECQ, means steaming the starting grains and Jiupei separately, and after the fermented grains are steam-distilled, no new grains are added, but the starter is added before the second fermentation. Finally, the second fermented grains are steamed and the residue is discarded) technique of the Light flavor type Baijiu.

Figure 6.1. Baofeng Baijiu from Henan Province.

In general, sorghum is brewed with barley, wheat and pea to produce the low-temperature Daqu by the QZ-QS technique, followed by fermentation in jar underground, a Zengtong distillation and an artificial blending procedure. It has the characteristics of 'pure aroma, soft sweetness, a dry and refreshing mouthfeel, and a long after-flavor'. In 1989, Baofeng Baijiu was awarded the title of National Famous Baijiu at the Fifth NAAC. In 2008, the traditional technique of brewing Baofeng Baijiu was listed in the second batch of national intangible cultural heritage. The registered trademark 'Baofeng Brand' was recognized as a 'China Time-Honored Brand' by the Ministry of Commerce in 2010.

74. Baiyunbian Baijiu

Baiyunbian Baijiu, shown in Figure 6.2, is a typical representative of the Strong-Sauce Mixed flavor type Baijiu. It is produced in Songzi City, Hubei Province, and is a product of Baiyunbian Group. Baiyunbian Baijiu is named from a poem of Bai Li, who is known as the poet-immortal. In 759 A.D., Bai Li with his brother Ye Li and his friend Zhiqiu Jia, visited Dongting Lake, went up the river and enjoyed local alcohol in Hukou (today's Songzi City). Bai improvised the following words: 'Clear and quite the water of South Lake in autumn night; flying, flying into deep sky if riding the flowing water; shinning, shinning moonlight from the Dongting Lake; celebrating the beautiful scene with Baiyunbian'. The Baiyunbian Baijiu

Figure 6.2. Baiyunbian Baijiu from Hubei Province.

production process combines the characteristics of the Sauce and Strong flavor types, starting in September every year, and ending in June the following year: using sorghum as a raw material, making Daqu at high and medium temperature only with wheat, three times of feeding, nine times of fermentation, eight times of picking up Baijiu, 10 times of operation, six times of accumulation at a high temperature, the seventh round operation carrying out the third feeding, adding medium-temperature Daqu and then directly putting into the pit, followed by a fermentation in the half-brick and half-mud pit, and classified distillation and storage in porcelain jars. The Baijiu body has the features of 'aromatic and elegance, harmony with Sauce and Strong flavor, a thick and rich mouthfeel, a sweet and refreshing taste, and a mellow and long aftertaste'. Baiyunbian Baijiu was awarded the title of National Quality Baijiu three times at the Third, Fourth and Fifth NAAC.

75. Dong Baijiu

Dong Baijiu, as shown in Figure 6.3, is a typical representative of the Herblike flavor type Baijiu. The origin of Dong Baijiu is in Donggongsi Town, Zunyi City, Guizhou Province. It is named for its location and is the product of Guizhou Dong Jiu Co., Ltd. The earliest recorded history of Dong Baijiu is the 'cellar Baijiu' brewed by local distilleries in 1937. Dong Baijiu was awarded the title of National Famous Baijiu four times in succession at the second, Third,

Figure 6.3. Dong Baijiu from Guizhou Province.

Fourth and Fifth NAAC, and in 2006, the Ministry of Science and Technology and the State Secret Bureau designated the production process of Dong Baijiu as a 'scientific and technological secret'. In Dong Baijiu processing, Daqu and Xiaoqu are used as saccharifying starters. There are 95 Chinese medicinal herbs in Xiaoqu, and 40 Chinese medicinal herbs are contained in Daqu. The pit mud is basic and the pit is sealed using coal. Daqu and Xiaoqu are used for fermentation separately, and the fermented grains are distilled together. The unique technique contributes to the flavor style of Dong Baijiu with the characteristics of 'three high and one low (high contents of ethyl butyrate, alcohol and total acids, and low content of ethyl lactate)', and endows the Baijiu with rich terpene flavor compounds. Dong Baijiu has the sensory characteristics of 'elegant ester fragrance, slightly comfortable medicinal flavor, beginning mellow and rich flavor, and a refreshing and long-lasting after-flavor and -taste'. In 2008, the Dong flavor type Baijiu was identified as a new flavor type of Baijiu by the local standard of Guizhou Province (DB52/T550-2008). The registered trademark 'Dong' was recognized as a 'China Time-Honored Brand' by the Chinese Ministry of Commerce in 2010.

76. Fen Baijiu

Fen Baijiu, also called Xinghuacun (Xinghua village) Baijiu, is regarded as a typical representative of the Light flavor type Baijiu and

Figure 6.4. Fen Baijiu from Shanxi Province.

as the 'soul' of Chinese Baijiu. Fen Baijiu, as Figure 6.4 shows, is produced in Xinghuacun Town, Fenyang, Shanxi Province. The history of Fen Baijiu is recorded in *The History of Northern Qi Dynasty* that in the Northern and Southern Dynasties, Fen Baijiu was the royal alcohol praised highly by the Emperor Wucheng of the Northern Qi Dynasty. In the Ming and Qing Dynasties, with the flow of migrants from Shanxi to the other regions of China and the rise of Shanxi merchants, Fen Baijiu was brought all over the country. Meanwhile, its traditional production process was further developed and modified to work better in different local conditions. For instance, underground jars were replaced by fermentation pits that led to the future development of different flavors of Chinese Baijiu. In the period of the Republic of China, Fen Baijiu made from sorghum was awarded the First-Class Grand Medal in the Panama World Fair in 1915. After the foundation of the People's Republic of China, the Ministry of Light Industry launched 3 pilot projects of Baijiu in 1964, in which the research of Fen Baijiu identified ethyl acetate as the major aroma component of Fen Baijiu. Fen Baijiu was awarded the title 'National Famous Baijiu' in the First to Fifth NAAC from 1952 to 1989. Fen Baijiu is made from the production process 'QZ-ECQ', that is, sorghum is used as the grain material and low-temperature Daqu made from barley and peas is used as the starter, which are then steamed separately, followed by fermentation in underground jars, distillation in Zengtong and blending. This production process pays attention 'clear to the end'. Thus, it tastes clear, mild, pure, soft and sweet with a long aftertaste. The Baijiu-making skills of Xinghuacun Fen Baijiu were enrolled in the 1st List of the National Non-Material Cultural Heritage in 2006. The registered trademark 'Xinghuacun' was certified as a 'China Time-Honored Brand' by the Ministry of Commerce in 2010.

77. Guojing Bandaojing Baijiu

Guojing Bandaojing Baijiu is produced by Shandong Bandaojing Co. Ltd. in Gaoqing County, Shandong Province. This company holds two major brands 'Guojing' and 'Bandaojing', as shown in Figure 6.5, and produces Baijiu of the light Strong flavor, multi-grain Sesame flavor, and Sesame & Sauce flavor types. It is a leading company of Sesame flavor Baijiu and a famous producer of the multi-grain Sesame flavor Baijiu in China.

Gaoqing County is the origin of Chinese alcohol culture and the hometown of Di Yi, the ancestor of alcohol. The story, 'Di Yi made alcohol', is a non-material cultural heritage of Shandong Province. The excavation of alcohol vessels made of bronze such as Gong, He, Zun and Yi from the sites of the Western Zhou Dynasty in Chenzhuang, Gaoqing, the original capital of Qi established by Jiang Taigong and one of the 10 New National Archaeological Discoveries

Figure 6.5. Guojing Baijiu (left) and Bandaojing Baijiu (right) made from Shandong Province.

in 2009, prove the long history of alcohol-making in the area where Guojing Baijiu and Bandaojing Baijiu are produced.

According to *The History of Gaoyuan*, 1000 years ago, Kuangyin Zhao, the First Emperor of the Song Dynasty, arrived at Gaoqing with his troops and drank the spring from Bandaojing (overturned well). So, the alcohol was named after this allusion. In the period of the late Ming Dynasty and early Qing Dynasty, there were seven large alcohol brewing workshops, Longxiang, Ruiqi, Hongchang, Dasheng, Guangji, Tianxiang and Jinyi, in the Bandaojing area. The alcohols were popular for their strong sesame flavor. In the Qing Dynasty, the local brewers added rice and millet in the ingredients, which along with the improved production process led to the more delicate and purer taste of the original alcohol. After the foundation of the People's Republic of China, the state-owned Gaoqing distillery was established based on the previous well pit factories. In 1998, Shandong Bandaojing Co. Ltd. was established.

The traditional Baijiu-making skills of Guojing Bandaojing, also called 'well pit technique', originated in the Song Dynasty and were enrolled in the List of Non-Material Cultural Heritage in Shandong in 2009. The site of the well pit of Bandaojing was certified as the cultural relic's conservation unit in Shandong in 2015. It takes five steps to make Jiuqu, the starter. Daqu and Fuqu are used properly. The raw materials contain six different grains that are piled up at high temperature. The solid-state fermentation is in the well pit at a high temperature. The raw grains and the fermented grains are steamed separately, and Jiupei is distilled in layers. The different qualities of fresh Baijiu are picked up and kept separately. After years of storage and blending, an excellent Baijiu is produced with its elegant flavor, smooth and harmonious mouthfeel, and a long and pleasant aftertaste.

Guojing Bandaojing is a typical Chinese famous Baijiu, a protected product of China Geographic Indications, and has been certified a 'China Time-Honored Brand' by the Ministry of Commerce. The 9th facility for the solid-state fermentation of grains was recorded in 'The Great World Guinness' in 2007. The Sesame flavor type and Strong flavor type Baijiu of Guojing Bandaojing received organic certification in 2011. The first academician workstation of the Baijiu industry was established in 2012. In 2014, the 1915 chateau of Guojing was recorded in 'The Great World Guinness' and became 'The No. 1 Chateau of Chinese Baijiu'. Bandaojing Baijiu became a recognized and protected product of China and Europe Geographic Indications in 2017.

78. Kweichow Moutai Baijiu

Moutai Baijiu (shown in Figure 6.6), produced in Moutai Town, Renhuai City, Zunyi City, Guizhou Province, is the origin of the Daqu Sauce flavor type Baijiu. The style has a 'prominent sauce flavor, which is elegant and graceful, delicacy, mellow and rich and full body, and lasting after-aroma from empty cups'. It has won the title of National Famous Baijiu from the First to Fifth NAAC.

According to *the Records of History*, a 'jujiangjiu (a type of alcohol made by Broussonetia papyrifera fruit)' was produced and popular in South Vietnam (in the area of Renhuai City today) in the sixth year of the Jianyuan period of the Western Han Dynasty (135 B.C.). After the Hongzhi reign in the Ming Dynasty, as the Shannxi salt merchants controlled salt in Sichuan and entered Guizhou, Moutai Town, an important wharf of Chishui river, gradually flourished, which promoted the development of the Baijiu industry in the region. Zhen Zheng, a poet of the Qing Dynasty, had stated the following sentence which reflected the situation: 'salt going from Sichun to Guizhou

Figure 6.6. Moutai Baijiu from Guizhou Province.

Province, traders gathering in Moutai from Qin area'. In the forty-third year of the Emperor Kangxi of the Qing Dynasty (1704 A.D.), 'Qisheng distillery' officially named the alcohol as Moutai Baijiu. According to the *Records of Old Zunyi Mansion* in the Qing Dynasty, during the Daoguang period (1821–1850 A.D.), there were at least 20 distilleries producing Moutai Baijiu and consuming more than 20,000 stones of grains on an annual basis. Stone is a unit of measurement, and one stone is approximately equivalent to 50 kg. These numbers suggested the prosperity of brewing industries in Moutai Town in the Qing Dynasty. In 1951, three private breweries, Chengyi, Ronghe and Hengxing, in Moutai Town were merged to establish a state-owned Moutai distillery, which is now converted into Kweichow Moutai Co., Ltd.

The traditional production of Moutai Baijiu uses the local glutinous red sorghum as a raw material, fermented in the pit with stones as walls and mud as the bottom. The final Moutai Baijiu is blended using the raw alcohols collected from different rounds, with different flavors, different alcoholic degrees and different storage years of Baijiu, after two times of feeding, nine times of cooking, eight times of fermentation, seven times of distillation, and long-term pottery jar storage. It takes more than four years for the raw material to be converted to the commercial product. The production process is characterized by 'three high, three low, three more, two long and one less'. The 'three high' refers to the high-temperature Daqu-making, high-temperature accumulation and high-temperature distillation. The 'three low' refers to low saccharification rate, low water content in the fermented grains and low Baijiu yield. The 'three more' refers to more grain consumption (2.5 kilograms of grains converted into 0.5 kilograms of Baijiu), more Jiuqu consumption (the ratio of Daqu to sorghum dosage is 1:1) and more fermentation rounds (eight rounds). 'Two long' refers to a long production cycle (about 10 months) and a long storage time (generally more than three years). 'One less' means a small amount of excipient consumption.

Kweichow Moutai Baijiu is highly praised by the Chinese people and plays an important role in China's political, diplomatic and economic life. Its brand value ranks at the top in the domestic Baijiu industries in China.

79. Gujinggong Baijiu

Gujinggong Baijiu, as shown in Figure 6.7, is one of the Eight Famous Baijiu products in China. It is produced in Gujing Town, Qiaocheng district, Bozhou City, Anhui Province. It is a type of Daqu Strong flavor Baijiu with a unique style of 'clear as a crystal, pure aroma like orchid, mellow entrance and lasting aftertaste'. Gujinggong Baijiu has been appraised as a National Famous Baijiu for four consecutive times, and is honored as a 'peony in Baijiu' and 'the top tribute in China'.

As recorded in *Qi Min Yao Shu (Important Arts for the Peoples Welfare)*, in the first year of the Jian'an period in the Eastern Han Dynasty (196 A.D.), Cao Cao dedicated the 'Jiuyunchun Baijiu' produced in Bozhou to Xie Liu, the Emperor Xian of the Han Dynasty. He also showed the *Jiuyunchun Baijiu Law*, the production technique of Jiuyunchun Baijiu, to Emperor Xian, which was greatly appreciated by the Emperor. Jiuyunchun Baijiu has become a kind of royal tribute alcohol since then, and the brewing workshops in Bozhou have been very prosperous, which contributes a lot to the development of the brewing industry in Bozhou. Anhui Gujing Gongjiu Distillery Co., Ltd. originated from the Gongxing tribute distiller in the 10th of Zhengde of the Ming Dynasty (1515 A.D.). In 1959, the state-owned

Figure 6.7. Gujinggong Baijiu from Anhui Province.

Bo Country Gujing Baijiu Factory was established, and in 1992, the group company was established.

On the basis of inheriting and developing the ancient brewing technology, Gujinggong Baijiu was fermented with Daqu of 'two flowers and one hot season', which has a storage period of not less than six months. The 'two flowers and one hot reason' refer to the production of a 'peach blossom Jiuqu' in spring, a 'hot season Jiuqu' in summer and a 'chrysanthemum Jiuqu' in autumn. The compositions of microorganisms and enzymes in Jiuqu produced in different seasons are different, so they have different saccharification and fermentation abilities. Gujinggong Baijiu is brewed using multi-Jiuqu. Three Jiuqu materials are mixed in different proportions and used in different fermentation rounds to optimize the performance of the different beneficial microorganisms in Jiuqu. With the unique techniques of 'three high and one low (high beginning starch content, high beginning acidity, high beginning Jiuqu content and low beginning temperature)' and 'three clear and one control (steaming raw materials, auxiliary materials and the bottom fermented grains in the pits, and controlling slurry and removing impurity)', the grains are fermented for 60 to 180 days, and they are collected by layers and the time of distillation, hierarchically stored in porcelain jars, and packaged after tasting, analysis, blending and aging. It takes at least five years from adding the raw materials to the production of commercial products. Gujinggong Aged Original is the current core product.

In 2010, the Gujinggong Baijiu Brewing Site was listed as a national key cultural relics protection unit and the 'Millennium Gujinggong Baijiu Traditional Brewing Techniques' was recognized as an intangible cultural heritage. In September 2018, the *Jiuyun Jiu Fa* was recorded in the Guinness World Records, as the earliest recorded method of brewing Baijiu.

80. Guilin Sanhua Baijiu

Guilin Sanhua Baijiu, as Figure 6.8 shows, is a typical representative of the Rice flavor type Baijiu. It is produced in Guilin City, Guangxi

Figure 6.8. Sanhua Baijiu from Guangxi Province.

Province, and is a product of Guilin Sanhua Co., Ltd. Guilin Sanhua Baijiu is also one of the 'three treasures of Guilin'. According to *Lingui County Records*, the industry of alcohol-making in Guilin has a history of more than 1,000 years. When Chengda Fan, a poet of the Song Dynasty, was an official in Guilin, he noted 'Sanhua Baijiu' in his *Guihai Yuheng Records* as 'coming to Guilin just like drinking good Baijiu'. In 1987, Pingwa Jia inscribed the inscription of Guilin Sanhua Baijiu: 'enjoy Guilin's landscape, drink Sanhua Baijiu, wish to live in the mountain city, and be a drunken fairy for a long time'. There are two stories about the name of Sanhua Baijiu. First, it is produced by steaming three times during brewing, and the shaking can produce numerous sparkling flowers; moreover, the Baijiu with good quality has fine hops and several layers, commonly known as 'three times making quality Baijiu'. Second, Sanhua Baijiu is brewed from pure Lijiang River water into 'Lishui flower'. High-quality rice is refined to brew 'rice flower', and the unique vanilla that grows in Guilin is made into Jiuqu called 'fragrant grass flower'. Together 'Lishui flower', 'rice flower' and 'fragrant grass flower' are called 'three types of flowers making the world aromatic'. Therefore, it is named Sanhua Baijiu. The typical production technique of Guilin Sanhua Baijiu is as follows: using rice as a raw material, Xiaoqu as a starter, through semi-solid fermentation (i.e. solid fermentation in the early stage, mainly for enlarged cultivation and saccharification, and liquid fermentation in the later stage), liquid distillation, a sealed ceramic vat, cave storage and artificial blending. Sanhua Baijiu has the typical flavor features with elegant and graceful honey flavor, sweet beginning taste, refreshing and long-lasting pleasant taste. The primary aroma components of Sanhua Baijiu are beta-phenylethanol and ethyl lactate. Guilin Sanhua Baijiu was awarded as a National Quality Baijiu at the Second, Third, Fourth and Fifth NAAC. In 2010, the registered trademark 'Guilin' was recognized as a 'China Time-Honored Brand' by the Chinese Ministry of Commerce.

81. Huanghelou Baijiu

Originating in Wuhan, Hubei Province, Huanghelou Baijiu is named after the Yellow Crane Tower, the 'first tower in the world'. The early product was the Daqu Light flavor type Baijiu, which was mild, mellow, elegant and graceful, with a refreshing and lasting after-flavor. In recent years, the Yellow Crane Tower brand has produced three flavor types of Baijiu products, including the Light, Strong, and Strong & Sauce flavor types.

Huanghelou Baijiu has an excellent reputation in history. *Records of the Weird* in the Southern Song Dynasty showed the origin of Yellow Crane Tower and Huanghelou Baijiu. It is said that a Baijiu seller, whose family name was Xin, was famous for brewing a unique pure, mild, clear and refreshing Baijiu. On the other side, a Taoist priest once drew a dancing yellow crane on the wall of Xin's tavern and left riding on the crane afterward. Therefore, the tower was built at the site and was called Yellow Crane Tower, and the Baijiu sold by Xin's tavern was named the Yellow Crane Tower Baijiu (Huanghelou Baijiu). In 1898, Zhidong Zhang, the governor of Hubei and Hunan Provinces, presented Huanghelou Baijiu to the Emperor Guangxu, and the Baijiu was given the name 'Tianchengfang' which indicated 'an excellent Baijiu made by the nature; the prosperous nation and strong people'. This was the predecessor of Huanghelou Baijiu.

After the foundation of the People's Republic of China, based on the 'Laotiancheng' Baijiu workshop, Wuhan State-Owned Distillery was established after consolidating other workshops such as 'Baikang' and 'Tongyuan'. In 1984, it was renamed Wuhan Yellow Crane

Tower Distillery and was restructured into Wuhan Tianlong Yellow Crane Tower Distillery Company Limited in 2003. In 2016, a strategic partnership with Gujinggong Group was established, and it was renamed Yellow Crane Tower Distillery Company limited in 2018. The 'Specialized Huanghelou Baijiu' produced by the Yellow Crane Tower Distillery was awarded the title 'China's National Famous Baijiu' twice by NAAC in 1984 and 1989, and became a famous Light flavor type Baijiu, which has a great reputation among the people of 'Huanghelou Baijiu in the south and Fenjiu Baijiu in the north'.

The Light flavor Baijiu, as shown in Figure 6.9, produced by the Yellow Crane Tower Distillery is made with quality sorghum as the raw material and Daqu as a sacchariferous and fermentative agent through the 'HZ-HS' technique, with one time feeding and twice fermentation. The Strong flavor type Baijiu in Yellow Crane Tower Distillery was brewed by the traditional 'LWZ' technique: five kinds of grains as the starting materials, medium- and high- temperature Daqu as the saccharifying and fermenting starter, and the 'HZ-HS' technique. The core products of the Yellow Crane Tower Distillery include the Big Light flavor type Baijiu series, the aging Baijiu series, the 'green' base Baijiu series, the tower series and the tiny Huanghelou series. Huanghelou Baijiu was awarded the title 'China Time-Honored Brand' by the Chinese Ministry of Commerce in 2011 and a 'Famous Trademark of China' by the Chinese Ministry of Commerce in 2017.

Figure 6.9. Huanghelou Baijiu from Hubei Province.

82. Hengshui Laobaigan Baijiu

Hengshui Laobaigan Baijiu, as shown in Figure 6.10, produced in Hebei Hengshui Laobaigan Liquor Co., Ltd., is a representative of the Laobaigan flavor type of Baijiu product. It has the distinctive characteristics of harmonious and mellow flavor and aroma, and long aftertaste.

Hengshui, known as Tao County and Tao City in ancient times, belongs to Jizhou and has a long history of brewing Baijiu. In the spring of the 16th year of the Eastern Han Dynasty and the Emperor Yongyuan (104 A.D.), Jizhou was banned from purchasing Baijiu because of heavy rain and the large amount of grain consumption for brewing Baijiu, which reflected the scale of Jizhou's Baijiu-making industry at that time. In the fourteenth year of Emperor Xuanzong's Kaiyuan era (726 A.D.), Zhihuan Wang, the great poet of the Tang Dynasty, was the governor of Taoxian County, Jizhou, and he praised Hengshui Baijiu as 'ten miles of Baijiu fragrant when opening of the jar, making a thousand drunken families'. During the Jiajing period of the Ming Dynasty, there were 18 famous distilleries in Hengshui. Among them, the Baijiu brewed by Deyuanyong was named Laobaigan because of its clean and high quality. 'Lao (old)' refers to a long history, 'bai (white)' refers to a clear body and 'gan (dry)' refers to its high alcohol content (up to 67 degrees) and no moisture residue after combustion. Baijiu produced in Hengshui was named after Laobaigan. In the Qing Dynasty, Hengshui Baijiu-making entered its heyday, with more than 30 distilleries in the city. In 1915, Hengshenghao distillery

Figure 6.10. Hengshui Laobaigan Baijiu from Hebei Province.

took the Hengshui Laobaigan Baijiu to the United States in the name of 'Zhili (official) Sorghum Baijiu' to attend the first Panamanian World Expo and won the first-class medal. After the liberation of Hengshui, in the spring of 1946, the Hengshui County government nationalized 18 private distilleries and established the local state-owned Hengshui Distillery of the Southern Hebei Administration. In November 1996, Hebei Hengshui Laobaigan Liquor Co., Ltd. was established.

Hengshui Laobaigan Baijiu is made with high-quality sorghum as raw material, and fermented with middle-temperature Jiuqu made by wheat in an underground jar for about 30 days. It is produced by 'Xucha Peiliao (XCA-PL, means that adding a certain amount of auxiliary material to the original fermented grains, and then mixing evenly and cooking)', 'HZ-HS' and 'LWZ' methods, segmented Baijiu picking, grading, pottery jar storage and blending.

In 2006, Hengshui Laobaigan Baijiu was recognized by the Chinese Ministry of Commerce as the first batch of the 'China Time-Honored Brand'. In 2008, the traditional brewing techniques of Hengshui Laobaigan Baijiu were recognized as a 'national intangible cultural heritage' by the Chinese Ministry of Culture.

83. Jiannanchun Baijiu

Jiannanchun Baijiu, shown in Figure 6.11, is a Strong flavor type Baijiu, and is produced in Mianzhu City, Sichuan Province. It is a

Figure 6.11. Jiannanchun Baijiu from Sichuan Province.

product of Mianzhu Jiannanchun Distillery Co., Ltd. of Sichuan. The brewing history of Jiannanchu Baijiu can be traced back to the Warring States Period. Eleven pieces of copper Baijiu-producing vessels such as copper barricade and Tiliang pot from the Warring States Period were unearthed in the Mianzhu area. The documentation of Jiannanchun Baijiu began in the Tang Dynasty. According to the *Supplement to the History of Tang Dynasty*, the national top Baijiu during the period of Tang Wude included the Baijiu fermented in 'the soil cellar of Xingyang' and the 'Jiannanshao' products. According to *Dezong Benji of the Tang Dynasty*, 'Jiannanshaochun Baijiu' in the fourteenth year of the Tang Dynasty was designated as a Royal tribute Baijiu. The discovery and excavation of the 'Tianyilaohao' distillery site systematically demonstrated a complete brewing process from raw material immersion and steaming to the final wastewater discharge. Jiannanchun Baijiu uses wheat to make Jiuqu, and is made from sorghum, rice, glutinous rice, wheat and maize by traditional techniques, such as solid fermentation in mud cellar, 'Xuzao Hunzheng (XZ-HZ, that is, the fermented grains and raw grain materials are mixed in proportion, then steamed together)', distillation using a Zengtong, and a pottery jar for storage, followed by artificial blending. It has the characteristics of 'one low, two long, three high, four appropriate and five fine works' (one low: low temperature in cellar; two long: long fermentation and storage periods; three high: high acidity, high residual lake and high Jiuqu temperature; four appropriate: appropriate moisture, temperature, ratio of grains to vinasse and husk filler; five fine processes: scientific stratification, uniform mixing, slow-fire distillation, picking up quality Baijiu and meticulous blending). The products of Jiannanchun Baijiu have the characteristics of 'aromatic, pure and elegant, soft and sweet, full and round, and harmonious aroma'. Jiannanchun Baijiu won the title of National Famous Baijiu three times in the Third, Fourth and Fifth NAAC. In 2005, the protection of geographical indication products for Jiannanchun Baijiu was officially stipulated by GB/T 19961-2005 promulgated by the Chinese General Administration of Quality Supervision and Inspection. The registered trademark 'Jiannanchun' was recognized by the Chinese Ministry of Commerce as a 'China Time-Honored Brand' in 2006. In 2008, the traditional brewing technique of distilled Baijiu-Jiannanchun Baijiu processing protocol was enrolled in the second batch of the national intangible cultural heritage list.

84. Jiugui Baijiu

Jiugui Baijiu is a typical representative of the Fuyu flavor type Baijiu. It is a product of Hunan Jiugui Liquor Co., Ltd. in the Jishou City, the Government of Tujia and Miao Autonomous Prefecture, Hunan Province, and is shown in Figure 6.12. The brewing history of Jiugui Baijiu is closely related to the customs of minority nationalities and the mysterious culture of western Hunan. According to textual research, the name of 'Drunken County' existed in the western Hunan in the Spring and Autumn Period, and the customs of drinking Baijiu when singing songs, with a barricade during seedling opening days, are prevalent even now. Jiugui Baijiu evolved from the Xiangquan Baijiu, which was finalized in the mid-1980s, designed by Mr. Yongyu Huang in 1988, and officially appeared in the market in 1989. Jiugui Baijiu, guided by Chinese culture, aroused great esteem in the art and cultural circles, and was established rapidly. Jiugui Baijiu was officially recognized as an innovative Fuyu flavor type Baijiu in 2005. The typical production technique of Jiugui Baijiu is as follows: sorghum is the main raw material for Baijiu-making, supplemented by rice, glutinous rice, wheat and maize, and followed by the steps of Xiaoqu bacteria culture and saccharification, fermentation of Daqu

Figure 6.12. Jiugui Baijiu from Hunan Province.

with grains, aroma upgrading in mud pits, steaming raw grains and fermented grains separately, being stored for aging in caves and combination blending. This unique technique endows Jiugui Baijiu with three basic flavor characteristics of Strong, Light and Sauce flavor types, as a sip with three flavors, a strong beginning, a light middle and a final sauce flavor. The Baijiu product has the unique flavor characteristics of elegant aroma, soft and sweet taste, mellow and delicate, pleasant aftertaste, rich aroma and refreshing body. In 2008, the protection of geographical indication products for Jiugui Baijiu was officially stipulated by GB/T 22736-2008 promulgated by the Chinese General Administration of Quality Supervision and Inspection.

85. Jingzhi Baijiu

Jingzhi Baijiu is a typical representative of the Sesame flavor type Baijiu. It is produced in Jingzhi Town, Anqiu City, Shandong Province, and is a product of Shandong Jingzhi Liquor Co., Ltd., as shown in Figure 6.13. According to *The Records of Anqiu*, in the Hongwu era of the Ming Dynasty, Anqiu paid an annual alcoholic beverages tax up to 'one hundred ingots and four strings' every year. In the Qing Dynasty, Baijiu-making in Jingzhi was flourishing. The Baijiu made in Jingzhi Town was the mellowest one in Anqiu according to *Anqiu Local Chronicle* in the Guangxu era. It was also recorded in *The General Records of Shandong* that Baijiu is flourishing in Jingzhi Town. During the period of the Republic of China era, there were 72

Figure 6.13. Jingzhi Baijiu from Shandong Province.

distillation pots in Jingzhi Town, and 'Taihe', 'Yuxing', 'Yushun' and a few other large-scale distilleries were formed. The production techniques of Jingzhi Baijiu combined the techniques for making Sauce, Strong and Light flavor types of Baijiu. Sorghum is used as a raw material, and Daqu and Fuqu are used as saccharifying starters, followed by a double-round fermentation, steaming and distillation separately, fermentation in a mud-bottom brick cellar and artificial blending. The production technique has the characteristics of high nitrogen ingredients, high-temperature accumulation, high-temperature fermentation and a long-term storage described as 'three high and one long'. Therefore, the Baijiu product has the typical style of elegant and prominent sesame flavor, mellow and delicate, harmonious flavor. The Sesame flavor type Baijiu is an innovative flavor type of Baijiu created after the founding of the People's Republic of China. In this process, Jingzhi Baijiu made outstanding contributions. In 1957, it was first found that there was a 'similar sesame flavor' in Jingzhi Baigan Baijiu. In 1965, the aroma components in the Sesame flavor type Baijiu were preliminarily studied by the Linyi pilot project of the Chinese Ministry of Light Industry. The industry standards of the Sesame flavor type Baijiu, QB/2187-1995, were issued by the former Light Industry Federation in 1995. In 2007, the National Standard of the Sesame flavor type Baijiu, GB/T 20824-2007, was promulgated by the Chinese General Administration of Quality Supervision and Inspection and the National Standardization Committee. In 2006, the registered trademark 'Jingzhi' was recognized by the Ministry of Commerce as a 'China Time-Honored Brand'. In 2008, the protection of Jingzhi Baijiu with geographical indications was officially stipulated by the GB/T 22735-2008 promulgated by the State Administration of Quality Supervision and Inspection.

The prototype of the 'Sanshilihong' Baijiu in the novel, film and TV drama of 'Red Sorghum' is the Jingzhi Baijiu.

86. Jinmen Sorghum Baijiu

Jinmen Sorghum Baijiu, as Figure 6.14 shows, made in Jinmen county, Taiwan Province, is a 'Jinmen flavor type' Baijiu which is a flavor type established by Jinmen Distillery Industrial Co., Ltd. In 1952, Jinmen

Figure 6.14. Jinmen Sorghum Baijiu from Taiwan Province.

Sorghum Baijiu was originally produced by Jiulongjiang Distillery established by General Lian Hu, the commander who garrisoned Jinmen at that time. Typically, Jinmen sorghum Baijiu is made with wheat and glutinous red sorghum as raw materials through a '3 highs, 2 lows and 1 turn-over' processing protocol, in the following manner: 'make the Jiuqu at a high temperature; steaming rice with a high pressure; collecting Baijiu at a high temperature', 'adding into the pit with a low temperature and fermentation at a low temperature' and 'turn over the fermented liquids'. The light flavor, pure sweetness and natural aroma of Jinmen Sorghum Baijiu also come from the unique storage in a tunnel. Jinmen Sorghum Baijiu is the No.1 Baijiu brand in Taiwan Province. As the symbol of Taiwan and the cross-strait relations, Jinmen Sorghum Baijiu, called 'Alcohol for Peace', is a good gift that Taiwan politicians bring when they have business visits. In 2005, when Zhan Lian, Chuyu Song and Muming Yu visited the Chinese mainland successively, they all brought Jinmen Sorghum Baijiu as a gift. In 2015, Yingjiu Ma presented Special Jinmen Sorghum Baijiu as a gift to President Jinping Xi. In 2007, Jinmen Sorghum Baijiu was issued the certificate of 'China Pure Grains Baijiu by Solid State Fermentation Method' by the China Food Industry Association.

87. Luzhou Laojiao Baijiu

Luzhou Laojiao, the origin of the national Strong flavor Baijiu products, is called 'ancestor of Strong flavor type Baijiu' and was created in Luzhou, Sichuan Province. Luzhou Laojiao Group is a large

Figure 6.15. Luzhou Laojiao Baijiu from Sichuan Province.

state-owned alcohol-making enterprise established on the basis of the 36 Baijiu workshops in the period of the Ming and Qing. The group owns many national famous Baijiu brands, e.g., 'Luzhou Laojiao' and 'Luzhou' Baijiu. 'Guojiao 1573' Baijiu, shown in Figure 6.15, is the representative of its high-end products, and the name comes from the earliest and the most complete '1573 pit groups' which have been continuously used since they were built without any interruption. The pits were awarded the title of 'National Priority Cultural Relic Protection Sites' for the first time by the Chinese Baijiu industry society in 1966. The traditional Baijiu-making technique of Luzhou Loajiao was formulated in 1324 A.D. and was appraised as the first batch of national intangible cultural heritage in 2006.

Luzhou Laojiao Baijiu was awarded the title 'China Famous Baijiu' in the First NAAC in 1952, and then continued to hold this title from the Second to the Fifth NAAC. In 1957, the Ministry of Light Industry checked and summarized the brewing techniques of Luzhou Laojiao Baijiu, and published the first textbook on brewing Baijiu named *Luzhou Laojiao Daqu Baijiu* in China. Since the 1960s, Luzhou Laojiao has opened a national training course on Baijiu-making technique and created the earliest Baijiu-making technical school in China, the Luzhou Laojiao Brewing Technical School. In 2013, 1619 pits of more than 100 years of age, 16 workshops of the Ming and Qing Dynasties and three Baijiu storage caves were recognized as Key Cultural Relic Protection Units in China. The number and category of cultural relics in Luzhou Laojiao rank at the top in

Baijiu industries. Luzhou Laojiao Baijiu has the typical style of colorless and transparent body, elegant cellar flavor, sweet and refreshing taste, soft and harmonious mouthfeel, and long-lasting aftertaste. The primary flavor component is ethyl hexanoate.

88. Lang Baijiu

Lang Baijiu, as Figure 6.16 shows, originated in Erlang Town, Gulin County, Luzhou City, Sichuan Province, and is produced by Sichuan Gulin Langjiu Distillery Co., Ltd. Lang Baijiu grows out of 'Huisha Langjiu' made by Huichuan, Jiyi and Xuzhi Distilleries during the reign of Emperor Guangxu in the Qing Dynasty. Lang Baijiu itself has no particular flavor style. Strong flavor, Sauce flavor and Mixed flavor types of Baijiu are all produced, but the Sauce flavor type Baijiu is the most typical product, which is made from sorghum and wheat through two additions of grains, nine times of steaming and heating, eight additions of Jiuqu, seven times of picking up liquids, storage in the cave and blending. The brewing characteristics of Lang Baijiu can be summarized as 'four highs, two longs, one large and several times', which means Jiuqu-making at a high temperature, accumulation at a high temperature, fermentation at a high temperature, distillation at a high temperature, a long production cycle, a long storage time, a large consumption of Jiuqu, and several times of fermentation and picking up liquids. Lang Baijiu possesses a prominent sauce aroma, elegant and graceful flavor, rich and full body and lasting after-aroma.

Figure 6.16. Lang Baijiu from Sichuan Province.

Lang Baijiu was awarded the title 'China Famous Baijiu' in the Fourth and Fifth NAAC. The registered trademark 'Lang' was confirmed as a 'China Time-Honored Brand' by the Chinese Ministry of Commerce in 2006. The traditional manufacturing technique of distilled Lang Baijiu was enrolled in the second List of National Non-Material Cultural Heritage in 2008.

89. Langyatai Baijiu

Langyatai Baijiu, a Strong flavor type Baijiu, as shown in Figure 6.17, produced by Langyatai Group Company located in Jiaonan, Qingdao, has the characteristics of the Strong flavor type Baijiu, summarized as a strong cellar aroma, a soft mouthfeel, a sweet beginning and a long aftertaste.

It is said that in the Western Zhou Dynasty, Ziya Jiang deployed the lord who was in charge of four seasons, spring, summer, autumn and winter, in Langyatai, which became the oldest observatory in China. In 472 B.C., Jian Gou, the king of Yue State, moved the capital to Langya and brought the alcohol-making skills from Wu and Yue to Langya. The people in Langya presented alcohol made from the spring in Langya to Jian Gou, who named the alcohol 'Langyahong'. According to *Historical Records*, after unifying the six kingdoms, the first emperor of the Qin visited the whole country five times and arrived at Langyatai three times. He gave the alcohol the name 'Langyatai Royal Alcohol'. Since then, many admirers such as ancient writers, scholars, calligraphers and chivalrous men came to

Figure 6.17. Langyatai Baijiu from Shandong Province.

climb the sacred mountain, drank Langyatai alcohol and wrote poems. The name of Langyatai Baijiu reflects a culture lasting thousands of years.

The local Baijiu workshops merged into the first distillery affiliated to the Commerce Bureau of Jiaonan in 1958. Qingdao Langyatai Group Co., Ltd. was established in 2002 and the Baijiu industry is always the pillar of the Group. Langyatai Baijiu-making has a unique process which includes making Jiuqu from three different grains (wheat, barley and pea) and producing Baijiu from five grains (sorghum, rice, glutinous rice, wheat and maize) through multiple rounds of fermentation (2–3 rounds), picking up liquids in layers (three layers) and a long storage time (over 3 years). In the meantime, the production of Langyatai Baijiu adopts the technique called '1 high, 1 low, and 2 appropriates' (addition into the pit at a high starch content and a low temperature, and appropriate water content and acidity), and '3 clear and 1 control' (steaming of raw materials, adjunct materials and fermentative materials at the bottom separately, and water control and impurity removal). The fermentation of alcoholic fermentative materials lasts 80–240 days.

The Baijiu-making protocol of Langyatai Baijiu was enrolled in the List of Non-Material Cultural Heritage of Qingdao in 2015. Langyatai Baijiu was confirmed as a 'Shandong Famous Brand', a 'Shandong Time-Honored Brand' and one of 'Ten Regional Advantageous Brands of Chinese Baijiu Industry', and became the designated Baijiu of the Shanghai Cooperation Organization Qingdao Summit in 2018. 'Langyatai' and 'Xiaolanggao' are its two famous Chinese trademarks with unique characteristics.

90. Maopu Buckwheat Baijiu

Maopu buckwheat Baijiu, as shown in Figure 6.18, is a healthy Baijiu produced by Hubei Jinpai Co., Ltd. It has the characteristics of light-yellow color, elegant buckwheat flavor, mellow and light taste, and sweet aftertaste.

Maopu buckwheat Baijiu is made from high-quality tartary buckwheat. The raw Baijiu is first produced by a new intelligent mechanical Xiaoqu brewing process and stored in a ceramic vat for more than

Figure 6.18. Maopu buckwheat Baijiu from Hubei Province.

three years. The raw buckwheat Baijiu is then blended with a high-quality glutinous sorghum Xiaoqu Baijiu, Daqu Strong flavor Baijiu and Sauce flavor Baijiu in a certain proportion to obtain the base Baijiu. The active ingredients in tartary buckwheat, kudzu root, medlar and hawthorn are extracted by alcohol, and then concentrated to yield the functional component extracting solution. Finally, Maopu buckwheat Baijiu is produced by combining the base Baijiu with the extracting solution of the functional components.

Maopu buckwheat Baijiu not only retains the taste and aroma of the traditional Baijiu but also contains functional components of tartary buckwheat. The test results showed that besides ethyl acetate, ethyl hexanoate, ethyl heptanoate, ethyl lactate, phenol, *p*-methyl phenol, 4-methyl guaiacol, 4-ethyl guaiacol, 2-methyl pyrazine, 2,3-dimethyl pyrazine, trimethyl pyrazine, tetramethyl pyrazine, acetic acid, butyric acid, hexanoic acid, heptanoic acid, octanoic acid and decanoic acid, Maopu Buckwheat Baijiu also contains high levels of health beneficial components, i.e. tartary buckwheat flavonoids and puerarin. The concentration of the tartary buckwheat flavonoids is above 50 mg/L and that of the puerarin is above 5 mg/L.

Maopu buckwheat Baijiu is a typical representative of the buckwheat flavor type Baijiu. Since its launch in 2013, it has become a new type of healthy Baijiu widely recognized by the consumers. It creates a new direction for the development of healthy Baijiu.

91. Niulanshan Erguotou Baijiu

Erguotou Baijiu is the first one named after the production technique. It originated from the Beijing 'Shaodaozi' process, which flourished in the Kang and Qian age of the Qing Dynasty, and has a long history of more than 300 years.

Niulanshan Erguotou Baijiu (shown in Figure 6.19) is the general term of the series of Erguotou. Niulanshan Baijiu is produced by Beijing Shunxin Agricultural Co., Ltd., which is located in Niulanshan, Shunyi District, Beijing. At present, Beijing Shunxin Agricultural Co., Ltd is one of the leading enterprises and Niulanshan is the leading brand in Erguotou Baijiu industries across China. The products have outstanding characteristics of light, refreshing, mellow and clear taste of Erguotou. The leading products consist of five series, i.e. classical Erguotou, traditional Erguotou, 100 years, Zhenniu and aging Erguotou. They are sold all over China and exported to many countries and regions overseas. Niulanshan Erguotou Baijiu is deeply favored by consumers.

Niulanshan Town is named after the Niulanshan Mountain. Its Baijiu culture can be traced back to the early Zhou Dynasty, three thousand years ago. Bringing together the spirits of a sweet well and Chaobai River, Erguotou Baijiu appeared. Over the past 300 years, due to the dedication to its high quality, Niulanshan Erguotou has gained a very good reputation. New chapters are created by the ancient methods, and they will flourish forever. According to the

Figure 6.19. Niulanshan Erguotou Baijiu from Beijing.

Records of Shunyi County in the reign of Kangxi, Huangjiu and Baijiu were famous products in the Niulanshan area. The *Records of Shunyi County* in the Republic of China era records that there were more than 100 Baijiu workers brewing the high-quality and tasty Baijiu, which was a special product of Peiping and being sold to the neighboring counties or Pingshi. The Baijiu products became very popular, among which Niulanshan Baijiu was the most famous Baijiu.

Niulanshan Erguotou Baijiu is made with sorghum as the raw material, with Jiuqu made by pea and barley as the saccharifying starter, and fermented in jars underground. It has the characteristics of harmonious, elegant, soft taste and pure aroma. The high quality of Niulanshan Erguotou Baijiu is mainly due to the continuous improvement and perseverance of traditional techniques over the centuries in addition to its unique natural environment and long history. According to the historical records, the traditional brewing techniques of Niulanshan Erguotou Baijiu originated from the late Ming and early Qing Dynasties, and matured in the mid-Qing Dynasty. It has a history of 300 years. Over the hundreds of years, the traditional brewing process of Niulanshan Erguotou Baijiu has been continuously improved to perfection through innovations from one generation to another. It was officially listed as the national intangible cultural heritage in 2008. The inheritance and perseverance of traditional brewing techniques are the key attributes for Erguotou Baijiu becoming a high-quality Baijiu.

92. Quanxing Daqu Baijiu

Quanxing Daqu Baijiu (shown in Figure 6.20), a Strong flavor type Baijiu, is made in Chengdu, Sichuan Province. It is produced by Sichuan Quanxing Distillery Co., Ltd. The history of Quanxing Daqu Baijiu can date back to the period of the late Yuan and early Ming when it was called 'Jinjiangchun Baijiu'. The distillery site of Shuijingfang discovered by the archaeologists in 1998 further proved that Quanxing Daqu Baijiu was flourishing in the period of the late Yuan and early Ming. In 1824, 'Fushengquan', the time-honored distillery in Chengdu was renamed 'Quanxingcheng' and sold Baijiu

Figure 6.20. Quanxing Daqu Baijiu from Sichuan Province.

called ‘Quanxingjiu’. In 1950, Western Sichuan Monopoly Bureau bought ‘Quanxing Old Brand’ and produced Baijiu called ‘Quanxing Daqu Baijiu’. Quanxing Daqu Baijiu was awarded the title of ‘National Famous Baijiu’ three times at the Second, Fourth and Fifth NAAC. Quanxing Daqu Baijiu is made from sorghum. The medium-temperature Daqu is fermented from wheat. The process of Quanxing Daqu Baijiu adapts the traditional old cellar-laying techniques, along with solid-state fermentation, mixing the distillers’ grains with the raw grains, the HZ-HS technique, distillation at medium temperature, storage in jars and blending. Quanxing Daqu Baijiu has a strong cellar aroma, mellow and harmonious taste, soft and sweet feeling, and refreshing palate.

Shuijingfang Baijiu originated from Quanxing Daqu Baijiu. In 2002, GB18624-2002 issued by the General Administration of Quality Supervision, Inspection and Quarantine of the People’s Republic of China, officially regulated the geographical indication protection of Shuijingfang Baijiu. In 2006, the registered trademark ‘Quanxing’ was designated as a ‘China Time-Honored Brand’ by the Ministry of Commerce.

93. Shuanggou Daqu Baijiu

Shuanggou Daqu Baijiu, as shown in Figure 6.21, is a type of Strong flavor Baijiu produced in Shuanggou Town, Suqian City, Jiangsu

Figure 6.21. Shuanggou Daqu Baijiu from Jiangsu Province.

Province. It is a product of Jiangsu Baijiu Group. The verifiable predecessor of Shuanggou Daqu Baijiu was the Baijiu produced by Quande distillery in 1719, which is the fifty-eighth year of Kangxi in the Qing Dynasty. Shuanggou Daqu Baijiu was recommended to participate in the Nanyang Counseling Fair in 1910 and was awarded the first place among the famous types of Baijiu. Mr. Yat-sen Sun inscribed 'Shuanggou liquan' for Shuanggou Daqu Baijiu in 1912. Quande Zaofang hosted the Communist Party leaders from Huazhong and Huaibei many times during the War of Resistance against Japan. It was well known as the 'Anti-Japanese Hotel', and was highly praised by Yi Chen, the commander of the New Fourth Army. Shuanggou Daqu Baijiu is made with sorghum as the raw material, and high-temperature Daqu made from wheat, barley and pea as the saccharifying starter. It is brewed by the 'HZ-HS' technique, fermented at a moderate temperature in an old cellar, distilled in a double-bottom barrel steamer (Zongtong), with the liquids picked up fractionally and stored separately based on the grade. Shuanggou Daqu Baijiu has the characteristics of 'strong and elegant cellar aroma; mellow, sweet and soft taste; full and harmonious body; and long lingering and refreshing aftertaste'. Shuanggou Daqu Baijiu was awarded the title of the National Famous Baijiu consecutively at the Fourth and Fifth NAAC. In 2010, the registered trademark 'Shuanggou' was recognized as a 'China Time-Honored Brand' by the Ministry of Commerce.

94. Songhe Baijiu

Songhe Baijiu is a Strong flavor type Baijiu, which is produced in Zaoji Town, Luyi County, Zhoukou City, Henan Province, and is shown in Figure 6.22. It is the product of Henan Songhe Wine-marketing Industry Co., Ltd. Songhe Baijiu originated from Zaoji Baijiu. According to the legend, Emperor Xuanzong of the Tang visited Laozi, Er Li, in Luyi with Zaoji Baijiu as a gift in 743 A.D. According to the *Records of Luyi County* of the Qing Dynasty, 'millet was used for Baijiu-making by the local people' and 'sorghum was used during Baijiu-making process, which was named steaming Baijiu'. It was indicated that the Luyi area already had a large scale of Baijiu-making at that time. In 1968, more than 20 local distilleries were consolidated into the Luyi Baijiu factory. Since the water used for brewing was obtained from the Song River, the product was named 'Songhe Baijiu'. The typical production techniques of Songhe Baijiu are as following: use sorghum, rice, glutinous rice and maize as the raw materials, and a medium-high temperature Daqu made from wheat as the saccharifying starter; follow the traditional 'LWZ' process, followed by the 'XCA-PL' technique and the 'HZ-HS' method; harvest separately based on the quality, and blend with the flavoring liquor made by turning over and accumulating the grains and other special techniques. The Baijiu product has the characteristics of 'elegant cellar aroma, soft, sweet and refreshing taste, balanced fragrant aroma, and a long lingering taste'. Song River Baijiu was awarded the

Figure 6.22. Songhe Baijiu from Henan Province.

title of the National Famous Baijiu in 1989 at the Fifth NAAC. In 2010, the registered trademark 'Songhe' was recognized as a 'China Time-Honored Brand' by the Ministry of Commerce.

95. Si'te Baijiu

Si'te Baijiu, a typical representative of the Te flavor type of Baijiu, is produced by Jiangxi Si'te Baijiu Co., Ltd. in Zhangshu Town, Yichun City, Jiangxi Province. The Si'te Baijiu product is shown in Figure 6.23. The verifiable history of Si'te Baijiu can date back to the late Tang Dynasty when 'Si'te Tushao' (also called 'Qingjiang Tushao') had been listed in the records. In the Yuan and Ming Dynasties, Zhangshu Town was famous throughout the country for its pharmaceutical industry and distillery. Its name is closely related to a distillery called 'Louyuanlong' during the reign of Emperor Guangxu in the Qing Dynasty. To distinguish its products from the Baijiu products made by the other distilleries, four Chinese characters 'Te' were glued on the jars of 'Louyuanlong'. The trademark 'Wangjinlou' was registered in 1958 and the trademark 'Si'te' was registered in 1982. Si'te Baijiu is made from whole rice and Daqu, which is made from flour, wheat bran and distillers' grains. It is fermented in a red stripped stone cellar. The overall style of the Te flavor type Baijiu is summarized as having 'all three types but not really one of them'. The unique process creates a special Te flavor product, which is characterized as a crystal clear, elegant aroma, pure taste and

Figure 6.23. Si'te Baijiu from Jiangxi Province.

soft body Baijiu. Si'te Baijiu was designated as the Te flavor type Baijiu in 1988 and was awarded the title 'National Quality Baijiu' in the Fifth NAAC. The National Standard of the Te flavor type Baijiu, GB/T 20823-2007, was published by the General Administration of Quality Supervision, Inspection and Quarantine of the People's Republic of China and the Standardization Administration of China in 2007.

96. Tianyoude Highland Barley Baijiu

Highland barley is the only grain that can survive and grow on a large scale at a high altitude. Highland barley Baijiu, as its name implies, is brewed using highland barley as the raw material, which is very unique compared to the general Baijiu. Tianyoude Highland Barley Baijiu, as shown in Figure 6.24, is a representative of the highland barley Baijiu from the Qinghai-Tibet Plateau area. It has the characteristics of the Light flavor type Baijiu. The unique raw materials of highland barley also contribute to the unique style of the highland barley Baijiu, which has the characteristics of elegant, pure, sweet, refreshing, balanced and harmonious aroma, and pleasant lingering taste.

According to the records of the *General History of Qinghai Province*, in the Yuan Dynasty (1264 A.D.), the Tu Nationality ancestors who lived in Huzhu County used boiled highland barley as the raw material, mixed with Jiuqu made from local herbs, and distilled

Figure 6.24. Tianyoude Highland Barley Baijiu from Qinghai Province.

to make a type of Baijiu known as Mingliu Baijiu. This type of Baijiu has a slightly turbid color, 30–40% of alcohol and a mild strength. The Tu Nationality ancestors liked this type of Baijiu. At that time, besides the self-sufficient family Baijiu-making, some small-scale Baijiu-making workshops appeared in the eastern agricultural region of Qinghai Province. By the 14th year of Jiaqing in the Ming Dynasty (1535 A.D.), there were more than 30 shops and 11 breweries in Weiyuan Town, Qinghai Province (now Huzhu County). Tianyoude Baijiu Distillery is the largest and most famous one. By the Qing Dynasty, there were many distilleries around Zhonggulou in Weiyuan Town, among which Tianyoude, Yongqinghe, Shiyide, Wenheyong, Wenyuhe, Chengyuan, Yixingde and Liuhening were the most famous breweries. In 1952, the Huzhu County People's Government consolidated the above eight workshops and established the state-owned Huzhu Distillery. In 1992, it was renamed Qinghai Highland Barley Distillery.

Highland barley planted 2,700 meters above the sea level on the Qinghai-Tibet Plateau is the main raw material for Jiuqu- and Baijiu-making in Tianyoude. This unique technique of using the same raw material for both Jiuqu and Baijiu is very rare in Baijiu-making processes in China. Normally, sorghum is the raw material for most of the Chinese Baijiu, and the raw materials for Jiuqu-making are mainly wheat, barley, bran, etc. Tianyoude Highland Barley Baijiu is fermented in cellars built from the granite strips. The granite is hard and corrosion resistant. It has good balancing and regulating effects on the temperature during the fermentation process. The floor of the cellar is made from the Qilian Mountain pine, which contacts with the Jiupei during fermentation to develop the unique flavor characteristics of the highland barley Baijiu. Combining the modern brewing techniques with the traditional process of 'Qingzheng Qingshao Siciqing' (QZ-QS-SCQ, means steaming grains and fermented grains separately and after the fermented grains are steam-distilled, no new grains are added, but the starter is added before the fourth time of fermentation), Tianyoude Highland Barley Baijiu formulates the different process conditions according to the 24 solar terms in the four seasons of a year. It feeds the raw materials four times a year, distills

16 times a year and brews continuously for 365 days. Twelve kinds of base Baijiu are obtained in each season and then totally 48 kinds of base Baijiu are produced in four seasons. After the natural curing process, the base Baijius are blended to generate a unique product.

In 2003, Tianyoude Highland Barley Baijiu was approved by the State Administration of Quality Supervision as 'the product protected by the geographical indications of the People's Republic of China' for its unique geographical environment, distinct raw materials, unique ingredients of Daqu, distinct brewing technique, unique fermentation equipment and unique product style. In 2009, Tianyoude Highland Barley Baijiu was recognized by the China Alcohol Industry Association as the representative of 'Chinese Light Flavor Type Baijiu (Highland Barley Raw Materials)'.

The production base of Tianyoude Highland Barley Baijiu is located in Tu Nationality Autonomous County, Huzhu County, Qinghai Province, and Lhasa City, Tibet Autonomous Region.

97. Tuopai Qu Baijiu

Tuopai Qu Baijiu, as shown in Figure 6.25, is a typical representative of the Strong flavor type Baijiu. It is produced in Tuopai Town, Shehong County, Sichuan Province, and is the product of Sichuan Tuopai Shede Spirits Co., Ltd. According to the *Records of Sichuan Province*, the production history of Tuopai Baijiu can be traced back to the Tang Dynasty when Shehongchun Baijiu was produced. In the thirty-fifth year of the Republic of China era (1946), Tianqu Ma, a successful candidate in the imperial examinations at the provincial level, renamed 'Shehongchun Baijiu' to 'Tuopai Baijiu' according to the implied meaning of 'Tuo spring water brews Baijiu with high quality, and the brand is famous for centuries'. The typical production process of Tuopai Baijiu is described as follows. Sorghum, rice and glutinous rice are used as the raw materials, and the medium- and high-temperature Daqu made from wheat and barley is used as the saccharifying starter. Bacteria are cultured in the artificial pit mud and go through the natural pit circulation and double-round bottom fermentation. The 'Xuzao Hunzheng Hunshao' method is used

Figure 6.25. Tuopai Qu Baijiu from Sichuan Province.

(XZ-HZ-HS refers to the fermented grains and raw grain powder that are mixed in a certain proportion; the fresh Baijiu and the grain are steamed at the same time. After steaming, the fermented grains are spread to cool down, sprinkled with Jiuqu and put into the cellar for XCA fermentation. Because distillers' grains are used continuously, it is also called 'XZ fermentation'. The fermented grains (mother grains) can be continuously recycled for several years, and never be lost, so they are called 'ten thousand years distillers' grains'). The production process is completed with the pottery storage and the artificial blending. The Baijiu body has the typical characteristics of the strong cellar aroma, clear, sweet and refreshing taste, soft and mellow mouthfeel, and a long lingering taste, especially the sweet and pure taste. In 1989, Tuopai Qu Baijiu was awarded the title of the National Famous Baijiu in China at the Fifth NAAC. In 2006, the registered trademark 'Tuopai' was recognized by the Ministry of Commerce as the 'China Time-Honored Brand'. In 2008, the 'Traditional Brewing Skills of the Distilled Tuopai Baijiu' were listed in the second batch of the national intangible cultural heritage list. The protection of the geographical indication products for Tuopai Qu Baijiu by the official regulations of the State Administration of Quality Inspection, GB/T 21822-2008, was implemented in 2008.

98. Wuliangye Baijiu

Wuliangye Baijiu, as shown in Figure 6.26, made from sorghum, long-grain rice, glutinous rice, wheat and maize, is one of the old

Figure 6.26. Wuliangye Baijiu from Sichuan Province.

eight types of famous Baijiu in China and the representative of the Daqu Strong flavor type Baijiu. It is produced in Yibin, Sichuan Province. Awarded the title 'National Famous Baijiu' successively from the Second to Fifth NAAC, it is famous for lasting aroma, pure flavor, sweet and refreshing mouthfeel, and harmonious, well balanced and rich taste.

Qingjiu was made in Yibin during the pre-Qin period. 'Zajiu' was made from the wheat, highland barley and rice in the Northern Dynasties (420–589 A.D.). 'Chunjiu' was made from four kinds of grain in the Tang Dynasty. 'Yaozixue Qujiu', the most mature embryonic form of Wuliangye Baijiu, was made from soybean, rice, sorghum, glutinous rice and buckwheat by Yao's family distillery in the Song Dynasty. In the first year of Emperor Hongwu in the Ming Dynasty (1368), a man from Yibin whose family name was Chen developed a secret Baijiu-making recipe by mixing the five grains of buckwheat, millet, rice, glutinous rice and sorghum to make a Baijiu, which was called 'mixed grains Baijiu' at that time. In 1909, Huiquan Yang, a successful candidate in the imperial examination at the provincial level in the late Qing Dynasty, changed the name 'mixed grains Baijiu' to 'Wuliangye Baijiu'. In the period of the late Ming and early Qing, there were four distilleries and 12 pits in Yibin. Before the foundation of the People's Republic of China, there were 14 distilleries, such as Deshengfu, Tingyuelou and Lichuanyong, and 125 pits. In the early 1950s, 8 distilleries merged into Sichuan Yibin Distillery of China Monopoly Company, which was renamed Yibin Wuliangye Distillery in 1959 and restructured into Wuliagye Yibin Co., Ltd. in

1998. The company owns the aged pits that have been used continuously from the period of Emperor Hongwu in the Ming Dynasty up to now.

Wuliangye Baijiu is made using the five grains of sorghum, long-grain rice, glutinous rice, wheat and maize as the raw materials. The protruding 'swelling qu' is made from wheat at a medium and high temperature. The raw materials are added to the mud pit at a low temperature, and then fermented for 70 days. The production process includes gathering Jiupei by layers, putting fermented materials into the pit by layers, putting fermented materials into Zongtong based on the quality, extracting the liquid and combining the extracts into the jar based on the quality, aging for years and blending.

The Wuliangye brand was designated as the first batch of the 'China Time-Honored Brand' in 2006. The traditional manufacturing technique of Wuliangye Baijiu was formally included in the List of the National Intangible Cultural Heritage in 2008.

99. Wuling Baijiu

Wuling Baijiu, one of the typical representatives of the Sauce flavor type Baijiu, is produced by Hunan Wuling Spirits Co., Ltd. in Changde City, Hunan Province. Wuling Baijiu, as shown in Figure 6.27, is named after the city Changde, which was called Wuling in the olden times. The alcohol-making history of Changde can date back to the pre-Qin period when people had the tradition of

Figure 6.27. Wuling Baijiu from Hunan Province.

enjoying alcohol at a feast. In the period of the Five Dynasties, 'Cui's alcohol' was famous and *The History of Changde* recorded that the alcohol made along Wuling creek was the only one found on earth, not in the paradise. Baijiu-making in Wuling was booming in the Qing Dynasty, which was reflected in *Wuling Zhuzhici*. It recorded that every village made Baijiu. Wuling Baijiu is one of the three representatives of the Sauce flavor type Baijiu in China. Its reputation was once tied with Moutai Baijiu and Lang Baijiu. The production process of Wuling Baijiu follows the typical process of the Sauce flavor type Baijiu with features of 'four high and two long'. It is made from glutinous red sorghum and a high-temperature Jiuqu made from wheat. The fermentation is carried out in pits with stone walls and a muddy bottom. The production cycle lasts one year and it is made through two additions of grains, nine times of steaming, two additions of sorghum, eight times of fermentation and seven extractions of the liquids. It has a strong sauce flavor, slight caramelized aroma, sweet and smooth mouthfeel, and elegant taste. The products are classified into 'Shaojiang', 'Zhongjiang' and 'Shangjiang' series based on the quality. Wuling Baijiu was awarded the title of National Famous Baijiu in the Fifth NAAC in 1989.

100. Xifeng Baijiu

Xifeng Baijiu is the representative of the Feng flavor type Baijiu. It is produced in Liulin Town, Fengxiang County, Shanxi Province. It is the product of Shanxi Xifeng Liquor Group Co., Ltd., and it is shown in Figure 6.28. Xifeng Baijiu has evolved through several generations, from Qin Baijiu to Qinzhouchun Baijiu, to Liulin Baijiu, to Tuoquan Baijiu, to Fengxiang Baijiu and finally to Xifeng Baijiu. The evolving production history of Xifeng Baijiu reflects the development of Chinese civilization. However, the Feng flavor type is still a relatively young type of Baijiu. It was established in 1992, and was officially promulgated by the National Standardization Committee in 1994 in the form of GB/T14867-94, the *National Standard for Feng Flavor Type Baijiu*. The typical processing techniques of the Feng flavor type Baijiu are summarized as follows. Japonica sorghum is used as the raw

Figure 6.28. Xifeng Baijiu from Shanxi Province.

material, and medium- to high-temperature Daqu made from the barley and pea is used as the saccharifying starter. The process of making the Xifeng flavor type Baijiu includes 'XCA-PL', mud pit fermentation, 'HZ-HS', Jiuhai storage and blending. The production cycle is one year. The Baijiu product has the characteristics of mellow and rich flavor, sweet, refreshing, harmonious and balanced flavor, and a long lingering taste. The major aroma components in the Feng flavor type Baijiu are ethyl acetate and ethyl caproate. The most innovative process of the Feng flavor type Baijiu is using Jiuhai to store Baijiu. The so-called Jiuhai is a woven large basket with thorns, with the basket wall covered with hundreds of layers of linen paper and coated with pig blood and lime; then it is coated with a formulation of egg white, beeswax and cooked rapeseed oil in a certain proportion, and finally kept to dry naturally. Xifeng Baijiu was awarded the title of National Famous Baijiu in China at the First NAAC. In 2004, the protection of geographical indication products for Xifeng Baijiu, i.e. GB 19508-2004, was officially issued by the General Administration of Quality Supervision and Inspection. In 2006, the registered trademark 'Xifeng' was recognized as a 'China Time-Honored Brand' by the Ministry of Commerce.

101. Yanghe Daqu Baijiu

Yanghe Daqu Baijiu, as shown in Figure 6.29, a Strong flavor type Baijiu, is produced by Jiangsu Sujiu Group in Yanghe Town,

Figure 6.29. Yanghe Daqu Baijiu from Jiangsu Province.

Suqian, Jiangsu Province. The traceable history of Yanghe Daqu Baijiu goes back over 400 years. According to the *History of Siyang*, Ji Zou, a poet in the Ming Dynasty, wrote in his *Ode to Baiyang River* that 'the spring water in Baiyang River is clear; many admirers come to Baiyang River; Spring arrives in February and the willow trees begin to bud; people come and go year by year; those who stay are fond of Yanghe Daqu'. The poem revealed a flourishing Yanghe Baijiu industry at that time. Yanghe Daqu Baijiu was awarded the title of 'International Famous Alcohol' at the Nanyang International Famous Alcohol Conference in 1923. The unique process of Yanghe Daqu Baijiu is developed based on its special qu-making and artificial aged pit mud techniques. It is made using multiple grains and combining various processing techniques, through long fermentation and storage times, and final blending with various flavors to generate the final product of Yanghe Daqu Baijiu. Yanghe Daqu Baijiu has the unique characteristics of 'sweet initial mellow mouthfeel, unctuous texture, refreshing and clear taste, and long-lasting fragrance', which confirms the 'sweet, mellow, soft, clear and aromatic' features of the Strong flavor type Baijiu. Yanghe Daqu Baijiu was successively awarded the title of National Famous Baijiu in the Third, Fourth and Fifth NAAC. In 2008, it was officially issued the geographical indications protection for Yanghe Daqu Baijiu, GB/T 20046-2008, published by the General Administration of Quality Supervision, Inspection and Quarantine of the People's Republic of China. In 2010, the registered trademark 'Yanghe' was designated as a 'China Time-Honored Brand' by the Ministry of Commerce.

The Blue Classic Series is a typical product of Sujiu Group, which includes Ocean Blue Series, Sky Blue Series and Dream Blue Series.

102. Yubingshao Baijiu

Yubingshao Baijiu, shown in Figure 6.30, is commonly known as the Chi flavor type Baijiu. The production origin is concentrated in the Pearl River Delta of Guangdong Province. Because lard or fat pork looks like jade, and one gets a cool feeling when touching the lard ('jade' is a homophone of 'meat' in Cantonese), liquor soaked in pork fat is called 'Yubingshao (jade ice) Baijiu'. The typical production process uses rice as the raw material, with the big alcohol 'cake' fermented from rice, soybean, rice cake leaves and Xiaoqu as the saccharifying starter. Fermentation is carried out simultaneously with saccharification, followed by kettle distillation, soaking in aged meat and blending. The Baijiu product has the typical characteristics of crystal pure, mellow and smooth flavor with dominant Chi flavor, and a refreshing aftertaste. In 1996, the State Bureau of Technical Supervision issued the national standard GB/T16289-1996 for the Chi flavor type Baijiu. At present, the larger producers of Yubingshao Baijiu include Guangdong Shiwan Liquor Co., Ltd., Guangdong Shunde Liquor Co., Ltd. and Foshan Taiji Liquor Co., Ltd. Yubingshao Baijiu produced by Shiwan Liquor Co., Ltd. won the title

Figure 6.30. Yubingshao Baijiu from Guangdong Province.

of China National Quality Baijiu twice at the Fourth and Fifth NAAC. Foshan City was designated as the 'Chinese Chi Flavor Type Baijiu Industrial Base' in 2010 by the Association of the Light Industry and the China Brewery Industry Association.

103. Yingjiagong Baijiu

Yingjiagong Baijiu, as shown in Figure 6.31, produced in Fuziling Town, Huoshan County, Lu'an City, Anhui Province, is a Strong flavor type Baijiu fermented by Daqu. Located in the hinterland of Dabie Mountains, Huoshan Town is the National Ecological Demonstration County. Yingjiagong Baijiu relies on a natural process. Its product has the typical characteristics of crystal clear color, outstanding compound aroma from five grains, elegant pit aroma, mellow and plump mouthfeel, sweet and refreshing taste, and long lingering taste. Yingjiagong Baijiu has won the 'Design Award of Chinese Baijiu' five times since 2012.

According to the *History of Huoshan*, in 106 B.C. when Emperor Wu of the Han Dynasty travelled south to the area of Huoshan, the officials and local villagers came to the dock near Caofang Village in the west of the city to welcome the emperor. They presented the alcohol to the emperor and he was very pleased after drinking it. Yingjiagong Baijiu then earned its name. The state-owned Fuziling Distillery was established in Huoshan in 1955 and it was restructured and renamed to Anhui Yingjiagong Jiu Co., Ltd. in 1997.

Figure 6.31. Yingjiagong Baijiu from Anhui Province.

Making Yingjiagong Baijiu requires certain ecological quality ideas, which integrate the following six aspects: ecological producing areas, ecological springs, ecological distillation, ecological cycle, ecological cave storage and ecological consumption. Based on the unique ecological environment and quality springs, Yingjiagong Baijiu is made with five types of grains as raw materials including sorghum, rice, glutinous rice, wheat and maize, and medium-temperature swelling Qu, which has thin skin, thick core and full mycelium, as the saccharifyng agent. The fermentation is carried out in an old mud pit for 90 days following the 'high in and high out' processing techniques (i.e. high starch content and high acidity into pit; high starch content and high acidity out of pit). Its core product is the ecologically cave-stored product series, as shown in Figure 6.31.

Yingjiagong Baijiu was successively awarded the titles of 'Product of the National Geographical Indications Protection' and the 'China Time-Honored Brand'. The traditional manufacturing technique of Yingjiagong Baijiu was included in the 'List of National Intangible Cultural Heritage'. The Yingjia Distillery was confirmed as the national green factory by the Ministry of Industry and Information Technology.

104. Development Trend of Baijiu

The traditional Baijiu process includes using grains as the raw materials and using Jiuqu as the saccharification and fermentation starter, through procedures of solid-state fermentation, distillation in Zongtong, aging in porcelain jars and blending. The above process is the mainstream technique and the basis of making healthy and delicious Baijiu. The new product development of Baijiu should insist on using the inherited traditions and seeking innovations in the process of advancement. One should use the traditional inherited methods first and then innovate to promote advancement, and finally achieve unification of the traditional technique and modern production.

Modern science and technology should be used to reveal the mystery of the traditional Baijiu. The basic scientific questions, such as the types of microorganisms in Jiuqu and the pit mud, their metabolic mechanisms and metabolites, the mechanism of enzyme effects in the brewing process, the development of the various compounds in Baijiu-making process, and their contributions to the flavor and health effects of Baijiu, need to be explored. This will build a scientific foundation for Baijiu production evolving from the 'inevitable kingdom' to the 'free kingdom'.

The future development of Baijiu should encourage diversity. In order to realize the modernization and intelligent control of Baijiu production, the traditional Baijiu industry should be reformed with modern technology and upgraded equipment based on the unique process of various types of Baijiu. For example, the saccharification and fermentation processes should gradually move from the underground to above the ground, and the fermentation process should achieve automatic temperature control instead of using natural temperature control.

The new product development and production of Baijiu should focus on the dual objectives of flavor and health. Healthy Baijiu should be developed following the rules of 'seeking internally, fortifying externally, and enhancing naturally'. The Baijiu blending process should realize the leap from manual blending to computer-controlled blending to make Baijiu more delicious and healthier.

We should carry forward the excellent Chinese Baijiu culture to present Baijiu to the world, to make Baijiu a worldwide beverage and to realize the internationalization of the Baijiu market.

Chapter 7
Famous Huangjiu

105. Dai County Huangjiu

The history of alcohol-making in Daizhou, Dai County today, Xinzhou, Shanxi Province, is over 1000 years old. The alcohol-making techniques had been developed and improved around the Yangmingpu area in Dai County in the period of the Qing and Ming Dynasties. The ballad 'quality Huangjiu are from Shaoxing in the south and Dai County in the north' proves that the typical representative of Huangjiu from north China is Dai County Huangjiu, which has been recognized as one of the 'two great Huangjiu series' with the Shaoxing Huangjiu. Figure 7.1 shows the product of Dai County Huangjiu.

The special taste of Dai County Huangjiu, one of the representatives of Huangjiu from the north, comes from its raw materials and the manufacturing technique. The raw materials of Dai County Huangjiu are the unique local crops. For instance, the high quality of millet comes from the long growth period due to the large day and night temperature difference in Dai County. Water is the 'blood' of Huangjiu. The water used to make Dai County Huangjiu is from Futuo River that flows through the whole county. The two mountains located at the north and south of the county have multiple layers of sands and gravel beneath which serve as filters for the groundwater.

Figure 7.1. Dai County Huangjiu from Shanxi Province.

The deep groundwater is clear and sweet enough to drink directly. The manufacturing technique of Dai County Huangjiu seems simple, but the processes of Qu-making, fermentation, aging, boiling and stir-frying with caramel are all done by experienced masters through observing by eyes, touching by hands and smelling by nose. Some controlling skills for different climate conditions can only be obtained by experience, and can hardly be described in words. Some technical requirements and standards for quality control still cannot be developed as specific theoretical indicators, and the brewing process is operated according to experience. The manufacturing technique of Dai County Huangjiu was enrolled in the list of provincial-level Intangible Cultural Heritage in Shanxi Province in 2008.

106. Guyuelongshan Huangjiu

Guyuelongshan Huangjiu, shown in Figure 7.2, is a typical product of Shaoxing Huangjiu Group Co., Ltd., and is a representative of Chinese high-end Huangjiu. The famous distillery originated from the Shenyonghe Distillery in 1664, and has the longest history in Shaoxing Huangjiu industries. Zhejiang Guyuelongshan Shaoxing Huangjiu Group Co., Ltd. established by the group company was the first listed company among Chinese Huangjiu companies, and is devoted to revitalizing the national industry and spreading Huangjiu culture. The Group Co., Ltd. also owns the National Research Center of Huangjiu Engineering Technology and the China Huangjiu

Figure 7.2. Guyuelongshan Huangjiu from Zhejiang Province.

Museum, and is the inheritance base of the manufacturing technique of Shaoxing Huangjiu, a national intangible cultural heritage. At present, the company has two 'Chinese Famous Brands' and four 'China Time-Honored Brands'. 'Guyuelongshan' is the iconic brand of Chinese Huangjiu and the only Huangjiu brand selected in the Asia Top 500 Brands.

Guyuelongshan Huangjiu was selected to be a part of the menu of Beijing Olympics and became a specially selected type of alcohol for the Olympic Games in 2008. A jar of Guyuelongshan was collected to be kept permanently by the China National Pavilion during World Expo Shanghai in 2010. In 2015, a 20-year Guyuelongshan appeared at the state banquet for President Obama and President Jinping Xi in the White House and witnessed the Sino-U.S. friendship. Eight Guyuelongshan brands were selected to be a part of the designated alcohol for the G20 Hangzhou Summit in 2016. Guyuelongshan was the designated alcohol of the 2nd, 3rd and 4th World Internet Conferences.

107. Hepai Huangjiu

Hepai Huangjiu, as shown in Figure 7.3, is produced by Shanghai Jinfeng Alcohol Co., Ltd.

The brand of Hepai Huangjiu is an inheritance and innovation of the Chinese traditional culture of 'He' (harmony). The character 'He' fully reflects the mild and intellectual characteristics of Huangjiu and

Figure 7.3. Hepai Huangjiu from Shanghai.

the Chinese tradition of 'harmony is precious'. 'Drink Hepai Huangjiu and make reliable friends' expresses the desire of modern people for communication, true friendship and a good quality of life.

In traditional Chinese culture, 'He' stands for harmony and coordination. As the saying goes, harmony brings money and auspiciousness. Harmony is the priority when communicating and interacting with others. Mutual harmony and coordination help us to develop and maintain great friendships and networks, and promote the level of achievements in one's life. Hepai Huangjiu is mild, pure, sweet and fragrant, which shows a moderate temper just like the harmonious relationships between human beings. This is also the essence of the culture of Hepai Huangjiu.

108. Jimo Aged Huangjiu

Jimo Aged Huangjiu, as shown in Figure 7.4, produced by Shandong Jimo Huangjiu Distillery Co., Ltd., is one of the typical representatives of Huangjiu in northern China. It has a good reputation of being 'the northern ancestor of Huangjiu'. Its history can date back over 2000 years, and the official record originated in the Northern Song Dynasty.

Jimo Aged Huangjiu is made from millets, aged volt wheat as Jiuqu and Laoshan spring as the water following the ancient six steps of Huangjiu-making, including using enough millets, timely addition

Figure 7.4. Jimo Aged Huangjiu from Shandong Province.

of Qu, sweet spring, quality pottery, clean steaming and full aging. The brown red liquid tastes slightly bitter and the aroma is enduring.

The traditional ancient six steps of Huangjiu-making inherited by Shandong Jimo Huangjiu Distillery Co., Ltd. have been enrolled in the List of Provincial-level Intangible Cultural Heritage in Shandong Province. Jimo Aged Huangjiu was awarded the title 'China Time-honored Brand' in 2006 and 'Jimo' was certified as a 'Chinese Famous Trademark' in 2010.

109. Kuaijishan Huangjiu

Kuaijishan Huangjiu, shown in Figure 7.5, is produced by Kuijishan Shaoxing Wine Co., Ltd., which grew out of the 'Yunji Distillery' established in 1743. Awarded 'China Time-honored Brand', 'China Famous Trademark' and 'Product of National Geographical Indications Protection', it was listed on the Shanghai Stock Exchange on August 25th, 2014, and became the 3rd listed company of the Huangjiu industry in China. With an inheriting history going back thousands of years, Kuaijishan Huangjiu is made from white glutinous rice, wheat Qu and Jian Lake water following the manufacturing technique being used for hundreds of years.

In 1915, 'Yunji Distillery' won the first international gold prize for Shaoxing Huangjiu at the Panama Pacific International Exposition

Figure 7.5. Kuaijishan Huangjiu from Zhejiang Province.

held in San Francisco, United States of America, and has won gold prizes at home and abroad 15 times to date. Kuijishan Huangjiu has long been recognized as 'Oriental Ruby' or 'the Crown of Oriental Famous Alcohols' by the international community.

110. Longyan Chengang (LYCG) Huangjiu

LYCG Huangjiu, a time-honored brand from Fujian Province, is a typical representative of red Qu Huangjiu in Fujian (shown in Figure 7.6). It originated in Gutian, Shanghang, Longyan, Fujian Province, in 1796. It was named after the process of adding Xiaoqu rice alcohol into alcoholic fermentative materials twice, which makes the materials move up and down three times before eventually falling down to the bottom of jar, and collecting the clear liquid in the upper layer for aging in the jar.

LYCG Huangjiu is a sweet red Qu Huangjiu made from quality glutinous rice, red Qu, crushed Qu, white Qu, the local medicinal Qu containing over 30 traditional Chinese medicinal herbs and quality rice alcohol, which is blended into the fermentation mixture. The red brown liquid is clear with an amber luster, sweet and pure taste without a feeling of stickiness, and has a unique flavor. In 2011, the manufacturing technique of LYCG Huangjiu was enrolled in the List of Provincial-level Non-Material Cultural Heritage in Fujian Province. The No. 166 announcement released by the General

Figure 7.6. Longyan Chengang Huangjiu from Fujian Province.

Figure 7.7. Lanling Huangjiu from Shandong Province.

Administration of Quality Supervision, Inspection and Quarantine officially approved the geographical indications protection for LYCG Huangjiu.

111. Lanling Huangjiu

Lanling Huangjiu, which is shown in Figure 7.7, is produced by Shandong Lanling Rice Wine Co., Ltd., a well-known large-scale production and sales company based in Shandong Province, which holds the oldest and most number of fermentation pits of grain alcohol, located in Lanling Town, Lanling County, Linyi City, Shandong Province.

Lanling Huangjiu is an exotic flower in Chinese alcohol with a history of more than 3000 years and a deep cultural heritage. Bai Li, a poet in the Tang Dynasty, wrote in his *A GUEST'S SONG*, 'Fine Huangjiu of Lanling with the fragrance of tulip is brought, it glitters like amber in a jade pot', which was praise for the color, aroma and taste of Lanling Huangjiu.

As one of the typical representatives of Huangjiu in north China, Lanling Huangjiu won the gold medal in the Panama Pacific International Exposition held in San Francisco in 1915. Since then, Lanling Huangjiu has been among the most famous alcohols in the world.

112. Nverhong and Zhuangyuanhong Huangjiu

The famous Shaoxing Huadiao Huangjiu is also called Nver Huangjiu (daughter Huangjiu). Han Ji who came from Shangyu in the Jin Dynasty wrote in *Narration of Grass and Trees in the South* that Nver Huangjiu was a must-have for a rich family that had a newborn daughter or when a daughter was married.

It is said that in the Jin Dynasty, a tailor who lived in the east of Shaoxing was delighted and prepared many jars of high-quality Huangjiu to celebrate the arrival of his son after he had known that his wife was pregnant. However, his wife gave birth to a daughter and he buried those jars of Huangjiu underground near the osmanthus trees in the courtyard as he preferred a son and was not happy to have a daughter. After 18 years, his daughter grew up with both beauty and talent. He was so pleased that he betrothed his daughter to his favorite apprentice. On the wedding day, he recalled the aged Huangjiu buried underground for 18 years and dug them out to serve to the guests. The guests were surprised by the delicious aged Huangjiu and the scholars were excited and acclaimed that 'Nverhong is a great huangjiu that cultivates a wonderful daughter'.

After hearing the story, his neighborhood began to do what he had done, as shown in Figure 7.8: burying Huangjiu underground at the birth of a daughter and digging it out to entertain guests at her

Figure 7.8. People buried Huangjiu underground at the birth of a baby.

wedding ceremony. The story spread far and widely. The custom of preparing Nverhong when a daughter was born and saving it to treat guests on her wedding day was established. Afterward, someone buried Huangjiu underground when his son was born for possible future celebrations during his potential first place (a Zhuangyuan) in the imperial exam. This Huangjiu was named 'Zhuangyuanhong'. It needs to be pointed out that females were not allowed to take the imperial exam.

Also, the daughter of the tailor and her husband were both skillful and wore wedding clothes that they themselves made. A wedding is a happy event in one's life and the red color, which means good luck in Chinese culture, is an indispensable part of a wedding dress. The wedding dress worn by the groom was called 'Zhuangyuan dress' or 'Zhuangyuan dress in red color', while the dress worn by the bride was called 'Nver dress' or 'Nver dress in red color'. This is the origin for the fashionable dresses of 'Zhuangyuanhong' and 'Nverhong' today.

Since then, Shaoxing residents prepared several jars of high-quality Huangjiu when kids were born. The Huangjiu was called 'Nverhong' if a girl was born, while it was called 'Zhuangyaunhong' if a boy was born, and the respective Huangjiu products today are shown in Figures 7.9 and 7.10. Painters were hired to draw patterns of 'booming flowers with a full moon' or write characters of 'good luck and happiness' on the jars, which were sealed with soil paste and stored in an underground cellar. When their children grew up and got married, those Huangjiu jars were taken out for the wedding ceremony.

Figure 7.9. Nverhong Huangjiu from Zhejiang Province.

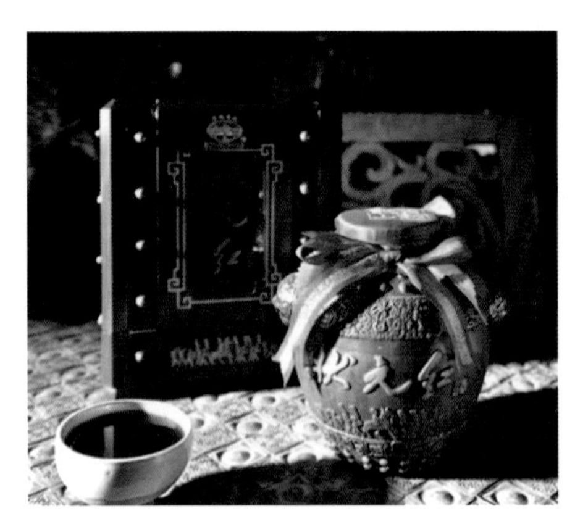

Figure 7.10. Zhuangyuanhong Huangjiu from Zhejiang Province.

The longer the Huangjiu is stored, the better quality and purer it will be. So, it is called 'aged Huangjiu'. Parents prepared aged Huangjiu for their children hoping that they will be well received like the aged Huangjiu, and will behave and interact well with others when they grow up. On the contrary, parents also hoped that their children can be like the aged Huangjiu through years of aging, more understanding of the world and more adept at dealing with people. Today, the whole process of this custom can hardly be found anymore. However, a good-quality Huangjiu is still an indispensable part of a wedding day in many regions.

113. Shaoxing Jiafan Huangjiu

Shaoxing Jiafan Huangjiu is a major product in the Huangjiu market, as shown in Figure 7.11. Jiafan Huangjiu, as the term suggests, means the addition of 'fan (rice)', that is, to use a greater amount of glutinous rice as the raw material.

Based on the manufacturing technique of Yuanhong Huangjiu, Shaoxing Jiafan Huangjiu is made by using an increased amount of glutinous rice and an improved processing protocol. Because of the high concentration of the fermented liquids in the fermentation mixture of Jiafan Huangjiu, and the high sugar and high alcohol contents of the finished product, the Huangjiu is mellow and smooth, commonly recognized as a 'thick body'. Shaoxing Jiafan Huangjiu is famous locally and internationally for its bright yellow and glossy color, rich aroma, fresh and pure taste, and appropriate sweetness.

Figure 7.11. Shaoxing Jiafan Huangjiu from Zhejiang Province.

The longer it is stored, the better it tastes. The long storage time has no negative impact on its quality.

114. Shikumen Huangjiu

Shikumen Huangjiu, as shown in Figure 7.12, is produced by Shanghai Jinfeng Huangjiu Co., Ltd., the biggest enterprise producing Huangjiu in Shanghai, which grew out of the Shanghai Fengjing Distillery established in 1939. 'Shikumen' was certified as a 'China Time-honored Brand' in 2008.

Shanghai, which used to be a paradise for adventurers, is a heaven for entrepreneurs now. Shikumen (Shiku gate) represents a typical style of architecture with unique features of Shanghai, and contains both Chinese and western style components. It has become the mark of cultural diversity in Shanghai and a gate connecting yesterday and today. Shikumen Huangjiu illustrates the special charm of Shanghai culture with the mixture of Chinese and western features, breaks the stereotype of traditional Huangjiu and creates a brand new Huangjiu culture. It has become the No. 1 brand in the high-class Huangjiu market in Shanghai and a famous brand in China. Obtaining the sponsorship for Huangjiu at Shanghai Expo in 2010, Shikumen Huangjiu, together with Shanghai, showed the world the charm of 'opening Shikumen to welcome guests from all over the world' with a smile in the most memorable Shanghai fashion.

Figure 7.12. Shikumen Huangjiu from Shanghai.

Figure 7.13. Shazhou Quality Huangjiu from Jiangsu Province.

115. Shazhou Huangjiu

Shazhou Huangjiu, shown in Figure 7.13, is produced by Jiangsu Zhangjiagang Brewery Co., Ltd., which originated in the reign of Emperor Guangxu. Growing out of the joint state–private ownership in 1956, it was renamed state-owned Shazhou Distillery in 1976 and the present name in 1999. It owns the brand series of Shazhou Youhuang Baijiu, Jiangnan Image Baijiu, Lucky Stars Baijiu, Stars of Taihu Baijiu and more than 150 products. Shazhou Quality Huangjiu is its most famous brand series.

Shazhou Quality Huangjiu was awarded a 'China Time-honored Brand' and enrolled in the first group of the List of Non-Material Cultural Heritage in Zhangjiagang City in 2005. It was awarded a 'China Famous Brand' in 2007 and a 'China Famous Trademark' in 2012. Shazhou Quality Huangjiu, typical of the Jiangsu style Huangjiu, is a Light flavor type Huangjiu.

116. Tapai Huangjiu

Tapai Huangjiu is produced by Zhejiang Tapai Brand Shaoxing Huangjiu Co., Ltd., which holds several honorary titles such as a 'China Famous Trademark', a 'China Famous Brand', a 'China Geographical Indications Product' and a 'Demonstration Base of Traditional Handmade Shaoxing Huangjiu'. Tapai Huangjiu was awarded a 'China Time-honored Brand' in 1999.

Figure 7.14. Tapai Huangjiu from Zhejiang Province.

Tapai Huangjiu, as shown in Figure 7.14, is manually processed. The production cycle lasts one year according to the solar terms. Qu is made in summer. The addition of raw materials and fermentation are done at the beginning of winter. Compression and picking up raw Huangjiu are carried out at the beginning of spring. The main product series include Original Huangjiu, Crude Huangjiu for Export, Handmade Huangjiu in Winter, Shaxoing Huadiao (Jiafan) Huangjiu, Lichun Huangjiu and Jiangnanhong Huangjiu. The representative product is the Tapai Original Huangjiu with a natural light-yellow colored liquid body.

117. The Development Trend of Huangjiu

With the economic development in China, the advancement of urbanization and the continuous increase of consumption, consumer desire has transitioned more and more from survival-oriented to health-oriented, enjoyment-oriented and diversity-oriented. Consumers are no longer satisfied with adequate amount of food, but pay more attention to tasty and healthy foods. Against the background of the 'Belt and Road Initiative' that emphasizes cultural communication and cooperation, Huangjiu, a symbol of traditional Chinese culture, has a bright future.

Huangjiu has become more and more popular for its low alcohol content, potential health benefit and good taste. Huangjiu producers should also pay more attention to the development trend focusing on flavor and health.

Trend 1: The functionality of Huangjiu becomes a major feature for the development of the Huangjiu industry.

The trend of multi-strain fermentation brings incomparable health beneficial features to Huangjiu. According to modern scientific research, Huangjiu is rich in health beneficial factors. For example, the functional oligosaccharide in Huangjiu can hardly be absorbed and generates no calories after entering one's body but promotes the growth of *Bifidobacterium*, a group of beneficial gut microbiota, improves the intestinal function and enhances immunity of the host. Huangjiu is rich in physiological active substances such as ligustrazine, ferulic acid and γ-aminobutyric acid, which act as nutraceuticals and have functions of eliminating free radicals, anti-oxidation, anti-aging, anti-thrombosis, anti-bacteria and inflammation, anti-tumor, reducing blood lipid and prevention of coronary heart disease. The lovastatin in red Qu Huangjiu may inhibit the activity of cholesterol biosynthesis. It is universally recognized as the preferred approach to treat hyperlipidemia, and to prevent and cure arteriosclerosis, coronary heart and cerebral vascular diseases. Therefore, to make Huangjiu rich in nutraceuticals is a major development trend for the Huangjiu industry in the future.

Trend 2: Advanced technology will be continuously applied in the Huangjiu industry.

The combination of scientific technology with novel high technologies such as the Internet, artificial intelligence and information, nanotechnology and new material technologies continuously results in innovations in Huangjiu production. IT, biological catalytic technology and biotransformation have been applied in raw material production, the fermentation process and consumption of Huangjiu. To enhance the level of specially designated substances or health factors in Huangjiu through precise control of fermentation, improve product competitiveness and the adaptability of Huangjiu enterprises, technology development and cooperative innovation become the priority.

Trend 3: Quality control is the first priority of Huangjiu enterprises and technical barriers should not be ignored.

During the inheritance and development, the diversity and differentiation in the Huangjiu industry development has to be taken into account. Traditional Huangjiu products are more classical (traditional skills, traditional flavor, outstanding function), while modern Huangjiu products (modern skills, no storage required, on-the-spot sale) are more fashionable. Furthermore, the problems and challenges of the Huangjiu industry have to be recognized. The major problem at present is the ease with which one can get drunk and the inconsistent quality. To improve the consumption experience, the inconsistent quality needs to be significantly improved. One possible solution may be the utilization of new technologies.

Chapter 8
Health Benefits of Baijiu & Huangjiu

118. Moderate Drinking Benefits to Human Health

The health benefits of moderate drinking are not only a conclusion of the thousands of years of drinking history but also evidenced by some modern scientific research. Traditional Chinese medicine believes that Baijiu can relax muscles, stimulate blood circulation, dispel dampness and prevent cold. As recorded by *Yellow Emperor's Internal Classic*, alcoholic drinks were used to treat diseases. *Hanshu Food and Goods Records* indicated that 'Alcohols is the beauty from the heaven, for emperors to take care of world, to pray for blessings, and to benefit human health'. 'Alcohol is the best medicine.' According to the *Compendium of Materia Medica*, moderate Baijiu drinking may 'eliminate cold, expel dampness and phlegm, modulate melancholy, and stop diarrhea'.

Recent studies have shown that mild to moderate alcohol drinking can reduce the risks of all-cause death and cardiovascular death, but heavy drinking can significantly increase the risks of all-cause death and cancer death. The subjects of the study were American adults. Researchers believed that mild to moderate drinking may have a protective effect on the cardiovascular system, while heavy drinking may lead to death. There is a delicate balance between the beneficial and harmful effects of ethanol, the main component of alcohol.

When evaluating the health benefits of moderate alcohol consumption, not only ethanol but also other minor components should be considered for their potential health benefits. The health benefits of grape wine drinking are well known because of its component resveratrol. The health benefits of Baijiu and Huangjiu may be more exciting because they contain more health factors compared with grape wines.

Baijiu and Huangjiu are rich in healthy components, which is the material basis of healthy Baijiu and Huangjiu. The diversity of flavor and healthy substances in Baijiu and Huangjiu is due to the diverse raw materials and brewing microorganisms, as well as the unique brewing techniques. So far, more than 2000 components have been detected in Baijiu, and more than 200 of those are beneficial to human health. For example, ethyl acetate is an anti-inflammatory and vasodilation chemical. Ethyl caproate may reduce lung inflammation and protect heart and lung functioning. Ethyl lactate has the functions of anti-inflammation and vasodilation. 4-Methyl guaiacol has the functions of promoting blood circulation and anti-aging. 4-Ethyl guaiacol may prevent disease and aging. Hexanoic acid, heptanoic acid, octanoic acid, decanoic acid, lauric acid, myristic acid, stearic acid, oleic acid, ethyl linoleate and ethyl linoleate all have the functions of inhibiting cholesterol synthesis. Tetramethylprazine has the functions of dilating blood vessels, improving microcirculation and inhibiting platelet accumulation. Most of the above active components can be found in Baijiu and Huangjiu. Huangjiu is a type of non-distilled alcoholic drink, and contains more non-volatile functional substances, such as polysaccharides and polypeptides.

Besides the potential health and hazardous effects on the human body, the influence of Baijiu or Huangjiu on the spiritual or non-physical aspects of the drinkers cannot be overlooked when discussing the relation between drinking Baijiu or Huangjiu and human health. Moderate drinking may relax the body and mind, stimulate the spirits and promote communication, so as to enhance friendships.

It must be pointed out that excessive drinking has the opposite effect. It is not only harmful to human health but also to social harmony. The proper amount of Baijiu or Huangjiu needs to be grasped by each individual. Generally speaking, the amount is good if one can have a good mood, a clear mind and use proper language after drinking. If one begins to talk too much, it is time to stop drinking.

119. Evolution of the Chinese Character 'Yi (Medical)'

Alcohol is closely related to medical topics and medicine in Chinese culture. As the sayings go, 'medicine comes from alcohols', 'alcohols and medicine come from the same origin' and 'alcohol is the best medicine'.

The *Yellow Emperor's Internal Classic* (a book about Chinese ancient medicine) says that 'alcoholic drinks are used to treat diseases'. In *Hanshu Food and Goods Records*, it is said that 'Baijiu is the beauty from the heaven, for emperors to take care of world, to pray for blessings, and to benefit human health'. In *Compendium of Materia Medica*, it is said that moderate Baijiu drinking can 'eliminate cold, expel dampness and phlegm, eliminate melancholy, and stop diarrhea'. The ancients realized the functions of alcohol in enhancing the effects of herbal medicines by promoting blood circulation, dispelling dampness, appetizing and invigorating the spleen, and improving liver function.

The common saying that medicine comes from alcohol can be observed in the two traditional Chinese characters meaning 'medical'. The traditional Chinese character '医 (Yi, means medical)' is first written as the character '毉 (Yi, means witch doctor)'. In ancient times, science and technology and productivity were not developed, and people could only pray for God's blessings when sick, which may be the origin of '毉 (Yi)'.

Later, the character '醫 (Yi, means doctor)' appeared. With the invention and drinking of alcohol, the medicinal value of alcohols was recognized, especially the alcoholic drinks with medicinal herbs that might have improved therapeutic effects. The traditional Chinese character '醫 (Yi)', with the bottom half presenting alcohol, was developed to reflect the medicinal importance of alcohol.

120. Medicinal Alcohols

Medicinal alcohols (medicinal Jiu) are made with Baijiu or Huangjiu, by extracting medicinal herbs soaked in the alcohol. Medicinal Jiu is a health product combining the benefits of alcohol and traditional Chinese medicines. It may modulate biological functions and improve human health. Figure 8.1 shows four famous brands of medicinal alcohol.

Medicinal Jiu has a long history, and the earliest extant Chinese medicine book *Huangdi Neijing-Su Wen* has recorded the treatment of diseases with medicinal Jiu. Zhongjing Zhang, a famous doctor of the Eastern Han Dynasty who was respected by later generations as a medical saint, recorded the methods of making medicinal Jiu, such as the red- and blue-flower alcohols and ephedrine alcoholic solution, in his book named *Summary of the Golden Plaque.* Simiao Sun, a medical doctor and a pharmacist in the Tang Dynasty respected by later generations as the king of medicines, comprehensively discussed the methods of making and taking medicinal Jiu in his book *Invaluable Prescriptions for Ready Reference.*

Shizhen Li, a famous medical scientist in the Ming Dynasty, listed 69 types of medicinal Jiu and their functions in the book named *Compendium of Materia Medica.* For examples, rehmannia glutinosa alcohol is good for tonifying weakness, strengthening the bones and musculature, promoting blood circulation, curing abdominal pain

Figure 8.1. Four famous products of medicinal Jiu.

and preventing hair whitening. Achyranthes bidentata alcohol can strengthen the bones and musculature, cure muscle atrophy, tonify weakness and eliminate chronic malaria. Angelica sinensis alcohol can promote blood circulation, strengthen the bones and musculature, relieve pains and regulate menses. Lycium barbarum alcohol can tonify weakness, benefit pneuma, dispel cold, strengthen yang-qi, stop tears and strengthen the waist and feet. Ginseng alcohol can invigorate the spleen-stomach and replenish 'qi', and tonify all weaknesses. Poria cocos alcohol is used to treat headache and dizziness, warm the waist and knee, and lighten diseases and pathogenic factors. Chrysanthemum alcohol is used to treat headache, improve chronic conditions of the ears and eyes, remove muscle atrophy and reduce the risk of all diseases. Huangjing alcohol can strengthen the bones and musculature, enrich essence, prevent hair whitening and treat all diseases. Bamboo leaf alcohol is good for curing different calenturas. Antler alcohol is used to treat impotence and weakness, frequent urination, fatigue and asthenia.

Medicinal Jiu is generally prepared by soaking, commonly known as soaking medicine alcohol. Ancient medicinal Jiu was mainly made with Huangjiu as the matrix, while modern medicinal Jiu is mainly made with Baijiu. The modern medicinal Jiu is no longer produced by the soaking method, but by using modern extraction technology to extract the active component of the medicinal materials and combining them into a base Baijiu according to scientific knowledge.

The modern medicinal Jiu belongs to the category of health food. Zhuyeqing Jiu, Jin Jiu, Huangjin Jiu, Baijin Jiu and Sanbian Jiu are several famous types of Chinese medicinal Jiu.

121. The Health Factors in Baijiu and Huangjiu — Alcohols, Acids and Esters

Alcohols, acids and esters are the primary minor components in Baijiu.

Inositol (cyclohexanol), mannitol (hexanol), sorbitol and other polyols are important components of sweetness and alcoholicity, and these compounds also have a variety of physiological effects. Inositol can be used to treat hepatitis and hypercholesterolemia. Mannitol has the effect of diuresis and may reduce the intraocular pressure. Sorbitol induces the secretion of the bile and pancreas, and prevents the rise of blood pressure and arteriosclerosis.

Low molecular organic acids, such as acetic, butyric and lactic acids, are important components in Baijiu. They are not only important flavor components but also the precursors of esters in Baijiu. Ethanoic acid, also known as acetic acid, is the primary acid component of vinegar, which has the function of dilating blood vessels and suppressing vascular sclerosis. Butyric acid can inhibit the growth and reproduction of cancer cells. Lactic acid is an essential organic acid for the human body, which promotes the growth of bifidobacteria and balances the microecology of the human body. Baijiu also contains long-chain fatty acids, which are beneficial to the human body, such as palmitic, linoleic and linolenic acids.

Esters are the most abundant minor components in Baijiu, which play important roles in Baijiu flavor, aroma, taste and characteristics. Ethyl acetate can accelerate the metabolism of inadaptable substances by enhancing renal function. Ethyl lactate can promote ethanol stimulation of cerebral cortex and induce excitement. Ethyl caproate can reduce lung inflammation and benefit the heart and lungs.

122. The Health Factors in Baijiu and Huangjiu — 4-Methyl Guaiacol and 4-Ethyl Guaiacol

4-Methyl guaiacol and 4-ethyl guaiacol are two important flavor components that have health effects in Baijiu, and are shown in Figure 8.2. 4-Methyl guaiacol, also known as 2-methoxy-4-methyl phenol, contributes soy sauce and smoked food flavors. 4-Ethyl guaiacol, also named 2-methoxy-4-ethyl phenol, contributes clove and smoke flavors. They are excellent free radical eliminators, and have antioxidant, anti-cancer, bacteriostasis, anti-infection and other beneficial effects and also enhance human immunity.

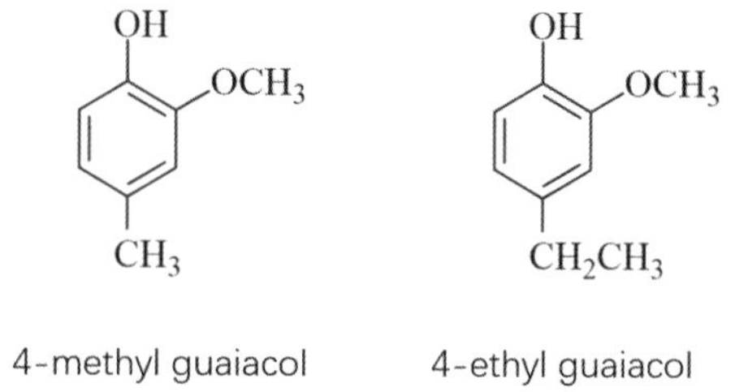

4-methyl guaiacol 4-ethyl guaiacol

Figure 8.2. The structure of 4-methyl guaiacol and 4-ethyl guaiacol.

Phenolic compounds cannot be synthesized in the human body. Food is the primary source of phenolic compounds. Ferulic acid is contained in wheat, corn, rice and other important raw materials for brewing Baijiu. Ferulic acid can be converted to other phenolic compounds such as 4-methyl guaiacol and 4-ethyl guaiacol by the microbial actions at a specific temperature and acidity.

4-Methyl guaiacol is a flavor component in the Sauce, Strong, Light, Medicinal, Mixed, Laobaigan and Sesame flavor types of Baijiu. Its concentration ranges from 15 to 1750 $\mu g/L$. 4-Ethyl guaiacol also exists in the Feng, Chi and Te flavor types of Baijiu with a level from 4 to 2390 $\mu g/L$. The contents of 4-methyl guaiacol and 4-ethyl guaiacol in the elegant Strong flavor type of Gujinggong Baijiu are relatively high. The concentration of 4-ethyl guaiacol in Huangjiu is greater than that in Baijiu with a concentration of 2500 to 7400 $\mu g/L$.

123. The Health Factors in Baijiu and Huangjiu — Ligustrazine

The scientific name of ligustrazine is tetramethylpyrazine. It is the primary active component of *Ligusticum chuanxiong hort* (shown in Figure 8.3), and is the metabolic product of *Bacillus subtilis*. According to the *Compendium of Materia Medica*, *Ligusticum wallichii* tastes pungent, and is warm natured. It targets potential problems in the liver and gallbladder, as well as the pericardium meridian. It

Figure 8.3. Traditional Chinese medicine, *Ligusticum chuanxiong hort.*

promotes blood and 'qi' circulation, dispels cold and relieves pain. Ligustrazine has the functions of scavenging free radicals, vasodilating, inhibiting platelet aggregation, improving microcirculation, protecting the liver and kidney, etc.

It was reported that ligustrazine was mainly produced by the non-enzymatic reaction of ammonia and acetoin, and was generated during the metabolism of functional strains in the Baijiu brewing processing. In addition, ligustrazine may also be produced by the Maillard reaction, thermal decomposition of proteins and thermal decomposition of amino acids during the process of Jiuqu-making and piling fermentation of grains. It has a unique aroma similar to coffee and nuts with a low threshold, and makes the Baijiu aroma elegant, full-bodied and soft, thus improving the quality of Baijiu.

Ligustrazine is present in the Strong, Sauce, Light, Medicinal, Mixed, Sesame and Laobaigan flavor types of Baijiu with a concentration range from 1 to 53020 μg/L. Ligustrazine has the highest content in the Sauce flavor type Baijiu and the lowest content in the Medicinal flavor type Baijiu. Research results showed that ligustrazine could enhance immune activity at a concentration of 0.10 μg/L, and its concentrations in the abovementioned flavor types of Baijiu are much higher than 0.10 μg/L. Therefore, its immune enhancement activity should not be underestimated.

Besides, ligustrazine is also an important trace component in Huangjiu. The concentration of ligustrazine in Huangjiu ranged from 3 to 73 μg/L.

124. The Health Factors in Baijiu and Huangjiu — Ferulic Acid

Ferulae Resina, shown in Figure 8.4, is a traditional Chinese medicine, which tastes spicy and is warm in nature. It has been recognized as having the potential to regulate the flow of vital energy, reduce swelling, invigorate blood circulation, relieve fatigue, remove phlegm and stimulate the nerves. Ferulic acid, 4-hydroxy-3-methoxycinnamic acid, is named for its widespread presence in the medicinal material named *Ferulae Resina*. Ferulic acid is also abundant in *Angelica sinensis*, *Ligusticum chuanxiong hort*, *Cimicifugae Rhizoma*, wheat straw, wheat bran, rice bran and corn husk. Wheat is used as the primary raw material for Jiuqu-making, whereas rice and corn are the important raw materials for Baijiu-making. These grains are important sources for ferulic acid in Baijiu.

Ferulic acid is a natural antioxidant and an internationally recognized anticancer substance in recent years. Ferulic acid can inhibit the synthesis of cholesterol in vivo, reduce blood lipids, inhibit platelet aggregation, effectively prevent thrombosis, reduce blood pressure and strengthen non-specific immunity. Additionally, it is a basic ingredient for producing pharmaceuticals capable of treating

Figure 8.4. Traditional Chinese medicine, *Ferula asafoetida*.

cardiovascular and cerebrovascular diseases and leukopenia, such as Xinxuekang and Limai capsule.

Ferulic acid is found both in Baijiu and Huangjiu. The ferulic acid content in Huangjiu is greater than that in Baijiu, generally at a range of 1560–2290 μg/L.

125. The Health Factors in Baijiu and Huangjiu — Polysaccharides

Polysaccharide is a group of carbohydrate polymers, which consists of more than ten monosaccharides linked by glycoside bonds. Polysaccharides are life essential substances, and are commonly water-insoluble amorphous solids, without sweetness, reducibility or optical rotation phenomena. Polysaccharides have many biological functions, such as immune modulating, anti-tumor, anti-virus, antioxidation, hypoglycemic, anti-coagulation, anti-ulcer, anti-radiation and anti-mutation activities.

The research suggested that polysaccharide had antioxidant activity when the concentration was greater than 1 mg/mL. Animal studies have shown that it could inhibit the growth and proliferation of cancer cells when the polysaccharide concentration of Huangjiu is greater than 25 mg/kg; a higher concentration was associated with a stronger inhibitory effect. Huangjiu is rich in polysaccharides, and the polysaccharide content is generally between 1 and 30 mg/mL.

126. The Health Factors in Baijiu and Huangjiu — Polypeptides

Polypeptide is a molecule with amino acids linked together by peptide bonds, and is an intermediate of protein hydrolysis. Compounds formed by dehydration and condensation of two amino acid molecules are dipeptides. Compounds formed by dehydration and condensation of three or more amino acid molecules are usually called polypeptides. Polypeptides may have blood pressure lowering, blood lipid lowering, cholesterol lowering, antimicrobial, antioxidant, anti-aging, anti-cancer, mineral absorption promoting, coagulation, nerve injury modulating and opioid-like antagonism effects.

Huangjiu contains polypeptides. According to previous studies, six polypeptides with the inhibitory activity of angiotensin-converting enzyme (ACE) have been detected in Huangjiu. ACE inhibitory peptides have become ideal targeted drugs for the potential treatment of hypertension, heart failure, diabetes mellitus with hypertension and other diseases. Forty-three bioactive peptides and three sensory active peptides were also reported in Huangjiu, some of which have the same amino acid sequences as the reported bioactive peptides.

In the process of Baijiu brewing, protein can be decomposed or converted into polypeptides by microorganisms. Baijiu contains a tripeptide, Pro-His-Pro, with ACE inhibitory and antioxidant activities, and a tetrapeptide, Ala-Lys-Arg-Ala, with antioxidant activity. However, most of the polypeptides are not distilled out but remain in the distillers' grains because of their high molecular weight.

127. The Health Factors in Baijiu and Huangjiu — Lovastatin

Lovastatin is a bioactive compound isolated from the fermentation products of *Monascus*. Lovastatin, a representative of the statins, can inhibit cholesterol biosynthesis, reduce serum cholesterol content, inhibit apolipoprotein synthesis in the liver and reduce the levels of low-density lipoprotein and triglyceride. Lovastatin is a primary clinic drug for reducing blood lipid.

Monascus is an important microorganism in Baijiu and Huangjiu brewing in China. Lovastatin is the secondary metabolite produced by *Monascus*. Clinical trials have shown that the synthesis of cholesterol in vivo may be suppressed at a lovastatin blood concentration of 0.001–0.005 μg/mL. Research also revealed that the lovastatin content in Baijiu ranges from 0.035 to 0.050 μg/mL, and 1–120 μg/mL in Huangjiu.

128. Healthy Drinking of Alcoholic Beverages

Alcohol acts as a double-edged sword. Moderate intake of alcohol is beneficial, while excessive intake is detrimental to human health. For the happiness and well-being of individuals, families and society, healthy drinking should be encouraged. The following five aspects are essential for healthy drinking of alcoholic beverages.

First, an appropriate amount of alcohol is important. One shall only have a moderate amount and avoid an excessive amount. Alcohol may promote interpersonal communication at a dining table. Alcohol may stimulate willingness of communication and bring about harmony in relationships. However, excessive drinking makes one brag and make thoughtless statements that easily hurt others' feelings. Drunkenness makes one to talk nonsense and is harmful to health. '100 mL of Baijiu makes one talkative; 250 mL of Baijiu makes one dare to brag; more than 250 mL may make one embarrassing; so moderate drinking is necessary for one's health.' The appropriate amount differs from one to another. The World Health Organization suggests that adults drink alcohol of no more than 25 g on a daily basis, which equals about 50 mL of Baijiu with a high alcohol content. Someone may get drunk at the first sip, while another may continue to drink even more than 250 mL of Baijiu without any problem. In general, it is appropriate to keep the intake in the 1/3 level of the 'bragging state'.

Second, it is important to drink with good manners, and not urge or encourage someone to overdose. Drinking with good manners promotes friendship, while reluctant drinking is counter-productive. Drinking manners reflect one's characters. As the saying goes in Reflections of *Book of Poetry*, 'Those who are grave and wise, in drinking won't get drank; but those who have dull eyes, in drinking will be sunk'.

Third, drinking should be at the right time, and minors should not drink alcohol. Many old sayings in China suggest drinking at the right time. For example, 'do not drink alcohols in the morning or to have a cup of tea in the evening'. In modern society, there are new regulations. Drunken driving is strictly prohibited and civil servants are not allowed to drink alcohol during work. As for the right time for drinking, our ancestors clearly stated that it is good for health to drink alcohol at the tenth of the twelve Earthly Branches, which is the time slot from 5 to 7 pm.

Fourth, do not try to dispel sadness with alcohol. Drowning sorrows in alcohol is a taboo in drinking. 'Draw a knife and cut the water, the water would nevertheless flow. Raise a cup to dispel your woe, woe comes after woe' (At a Farewell Banquet for *Yun Shu, the Official Editor, in Xie Tiao Tower, Xuanzhou* by Bai Li). 'Do not regard alcohols as water that may put out your worries. It more likes an oil poured into the fire that would make the smart smarter and the stupid stupider' (*Alcohols* by Qing Ai).

Fifth, encourage national alcohol drinking. Baijiu and Huangjiu, the essence of the five cereals, are the national alcohols that have been consumed by the Chinese for thousands of years and are rich in beneficial substances, the health factors. This is an incomparable point for other kinds of alcohol beverages.

Among the above five aspects of healthy alcohol drinking, amount control is a critical point. Several states of drinking alcohol are summarized: careful words, sweet words, thoughtless words and silence. Generally speaking, the amount of alcohol to bring one to the state of sweet words is already too much.

Chapter 9
Famous People & Alcohols

129. Confucius and Alcohol

Confucius (September 28th 551 B.C. to April 11th 479 B.C.) was a sage, but also a prodigious drinker, a great thinker and educator who has been respected as an exemplary educator and mentor for all ages throughout history.

A prodigious drinker loves alcohol and has his/her criteria for alcohol. Confucius only drank homemade alcohol, but not necessarily made by his own family, and did not drink the alcohol bought commercially. As the saying goes in *The Analects*, shown in Figure 9.1, 'do not eat the meat nor drink the alcohols bought from commercial'. Confucius lived in the Spring and Autumn period, 2500 years ago, when alcohols sold in the market possibly had problems of quality, so Confucius did not drink commercial alcohols.

Confucius was a gourmet who did not eat under eight circumstances: 'no rotted food, foods with a unpleasant color, the foods with a unpleasant smell, foods not fully cooked, the foods not for the season, foods not cut in a right way, the foods not with correct seasonings, and the non-home-made foods.' He preferred well-prepared delicate food and food with ginger and no overeating, and believed in the intake of adequate amount of meat, but not too much as compared to the intake of staple foods. However, Confucius did not set

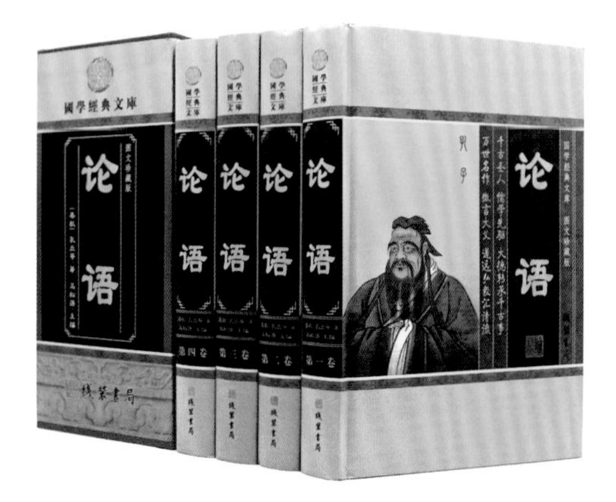

Figure 9.1. The Analects of Confucius.

restrict drinking and kept a moderate drinking habit with strong self-control. As the saying goes in *The Analects*, 'the intake of alcohols is not limited, but excessive drinking and behavior out of manners after drinking are prohibited'.

What deserves praise is that Confucius was 'never bothered or confused by the alcohols' (*The Analects*). Literally, 'never bothered or confused by the alcohols' means 'never puzzled, confused, or fettered by alcohols' and it is composed of two aspects. One is that it is not necessary that you have to go for a drink when you are invited. If it is not the right person who invites you or it is a party that does not fit your principles, you do not have to attend it. The other is that you do not have to drink when someone proposes a toast. If you cannot drink, just refuse it. The statement that 'distant relationship deserves a sip while close relationship deserves a swallow' is wrong.

Confucius' view of food and drinking may be shared and considered by the later generations.

130. Cao Cao and Alcohol

Cao Cao has been considered as a white-faced treacherous minister portrayed in Chinese opera and storytelling and works of fiction like *Three Kingdoms*. In fact, Cao Cao, a great politician and militarist in the period of the Three Kingdoms, was also an accomplished poet who wrote the well-known lines 'singing while enjoying Baijiu if you still can' and 'only Dukang may relieve my worries'. Cao Cao loved, enjoyed and made Baijiu. Cao Cao was closely tied to Baijiu and was respected as the God of Alcohol in Gujing by people from his hometown, Bozhou City.

According to *Qi Min Yao Shu (Important Arts for the Peoples Welfare)*, in the first year of Jian'an in the Western Han Dynasty (196 A.D.), Cao Cao presented the 'Jiuyunchunjiu' made in his hometown (Bozhou) to Xie Liu, Emperor Xian of the Han Dynasty, along with its manufacturing technique. Cao Cao wrote in his *The Methods to Make Jiutanjiu* that 'Zhi Guo, the former county magistrate from Nanyang, has Jiuyunchunjiu. 10 kg of Qu and 5 dans of water are used. Soak the Qu at the second day of the twelfth month of lunar year. When the ice melts at the first month of lunar year, the quality rice was selected for brewing after the dregs were filtered out. Rice was added every three days for a total nine times. I have learnt this way and it always works. The liquid is clear and the dregs are drinkable. If the alcohol from the nine-time rice addition still tastes bitter, a 10-time rice addition could improve the taste. I am honored to present it to you.' This is the documentary evidence of 'Jiuyunchunjiu' as tribute and the origin of Gujinggong Baijiu in Bozhou.

The Methods to Make Jiutanjiu by Cao Cao not only summarized the manufacturing technique of 'Jiuyunchunjiu' but also introduced the way to make the alcohol purer and stronger. This summary of Baijiu-brewing skills in Bozhou is similar to the current way of continuous addition of materials. So, some scholars think that 'after the period of Wei, the Baijiu-brewing ways of continuous addition of materials adopted by the makers came from the way reported by Cao Cao'. Cao Cao not only recorded the Baijiu-brewing skills but also improved them and made Baijiu by himself. He was a brewer and researcher for Baijiu-brewing. Surely, Cao Cao preferred drinking Baijiu and wrote many poems about alcohol and stories such as 'defining a hero through drinking alcohols', 'drinking alcohols while enjoying the river scene' and 'holding the spear and composing a poem'.

Cao Cao, the Baijiu lover, also prohibited Baijiu sometimes. In the twelfth year of Jian'an in the Eastern Han Dynasty (207 A.D.), the enduring famine resulted in a peasant uprising. 'The famine resulted in uprising. Cao Cao suggested prohibiting Baijiu. Rong Kong did not agree with him and used insulting words to dispute with him.' Although Cao Cao loved drinking Baijiu, he insisted on prohibiting Baijiu due to the national economy and the people's livelihood. This measure was to save grains and to protect agriculture and peasants when the famine was spreading. Xun Lu explained in his *The Relations between the Demeanor and Articles of Wei and Jin Dynasties, and Medicine and Alcohols* that Cao Cao liked drinking Baijiu, which was proved by his lines 'only Dukang may relieve my worries'. Why was his behavior contradictory to his opinions? Because he was a responsible man and he had to take action.

To commemorate Cao Cao's contribution to Baijiu, on September 19th of every year, a ceremony is held by Gujing residents to worship Cao Cao, the God of Alcohol, at the festival of Baijiu-brewing of Gujinggong Baijiu in autumn. The square of Baijiu God in Gujing, as

Figure 9.2. Square of Baijiu God in Gujing Town, Anhui Province.

shown in Figure 9.2, was built in Bozhou in 2015 where a 19.6-meter-high statue of Cao Cao holds a cup and sings before Baijiu.

131. Bai Li and Alcohol

Bai Li (701–762 A.D.), whose sculpture is shown in Figure 9.3, is the poet immortal and an alcohol immortal who wrote more than 1000 popular poems. The moon and alcohol are the most frequent words in his poems and even appear together in one poem many times, such as 'With a jug of wine among the flowers, I drink alone sans company. To the moon aloft I raise my cup, with my shadow to form a group of three' (*Drinking Alone under the Moon*); 'shinning, shinning moonlight from the Dongting Lake; celebrating the beautiful scene with Baiyunbian' (*Tour to Dongting Lake*); and 'I would when we sing holding our drink, the moon its beams to our beakers let fall' (*Holding Drink to Ask the Moon*).

Figure 9.3. The sculpture of Bai Li.

Bai Li loved and enjoyed drinking alcohol whenever he had alcohol and used to drink excessively in most cases. 'There are 36000 days in 100 years and 300 cups of alcohols should be drunk every day' (*Song of Xiangyang*) and 'Let mutton and beef be broiled for making merry, we should drain three hundred bumpers at one carouse' (*Carouse, Please*) are true portrayals of Bai Li's drinking.

Bai Li lived during the prosperous period of the Tang Dynasty with active external contacts when the alcohol-making industry was well developed and Baijiu, Huangjiu and grape wine were all produced and commercially available. Bai Li enjoyed different alcohols including Baijiu, Huangjiu and grape wine wherever he went.

The history of Chinese Baijiu dates back to the Western Han Dynasty. It is common that Baijiu was available in the Tang Dynasty. Bai Li had consumed Baijiu as evidenced by his poem 'Baijiu was made after I returned from mountains; the chicken grew up during the grain harvesting season; I asked an assistant to cook the chicken and prepare Baijiu; kids were smiling around' (*Farewell to Kids in Nanling and Heading to the Capital*). The reason Bai Li wrote 'cook chicken and prepare Baijiu' might be connected with traditional Chinese culture. As is known, '酉 (You, means alcohols)' means Baijiu. '酉 (You)' in the twelve earthly branches of the ancient Chinese calendar is matched with the rooster in the twelve animals of the Chinese Zodiac. This could be a coincidence or providence. To date, there are many dishes prepared from chicken and served with alcohol such as 'beggar's chicken', 'boiled chicken slices', 'Dezhou braised chicken' and 'red-cooked chicken, Daokou Style'.

Huangjiu was popular in the Tang Dynasty. Bai Li travelled across the country and what he had most was Huangjiu. He used to get drunk and forget where his hometown was. 'Fine huangjiu of Lanling with the fragrance of tulip is brought, it glitters like amber in a jade pot. If only the host can make the guest drunk, where home is he'd know not' (*A GUEST'S SONG*).

Han Wang (687–726 A.D.), a poet in the Tang Dynasty, lived in the period before Bai Li. The following lines in his *Song of Liangzhou* proved the existence of grape wine in the Tang Dynasty, even though it possibly came from the Western Regions: 'Fine grape wine in luminous cups of jade: to drink I want but the summoning Pi-pa on horseback is played. Laugh not, my dear friend, if I lie drunk on the battlefield. How many ever returned from battles anyway.' Bai Li also described grape wine in his lines. 'Duck heads green above Han water from far, as if the green grape wine fresh from brewing jar' (*Song of Xiangyang*) shows that Bai Li had known the process of brewing grape wine.

Although Li Bai wrote rationally, 'Draw a knife and cut the water, the water would nevertheless flow. Raise a cup to dispel your woe, woe comes after woe' (*At a Farewell Banquet for Shu Yun, the Official Editor in Xietiao Tower, Xuanzhou*), he still drowned his sorrows with alcohol. 'My mottled steed and the fur-lined robe of a thousand crowns. Let my boy lead and fetch out to barter for drinks divine, in order to banish with ye both our griefs eternal trine' (*Carouse, Please*) and 'Sorrows and depressions endless, three hundred cups of liquors in hands; too much unhappiness for little liquor, liquor down and sadness gone' (*Drinking Alone under Moon*) are the points of evidence.

Some dregs of feudalism are also in Bai Li's poems, among which the negative views account for a substantial part, for example, that life is a dream so we should enjoy the pleasure before it is too late and drink alcohol as much as we like. The following is another example: 'Seize the moments of content in life and make full mirth of them, let not your golden beakers stay empty to glint at the moon' (*Carouse, Please*). Bai Li was a victim of excessive drinking and one who drowned his sorrows with alcohol. It is said that he died from drowning after he got drunk and wanted to seize the moon's shadow reflected in the lake. This is a lesson that drinkers should learn.

132. Fu Du and Alcohol

Fu Du (712–770 A.D.) was called 'Poet Sage' and his poetry was called 'The History of Poems'. Fu Du, a great realistic poet living in the Tang Dynasty, was 11 years younger than Bai Li and they were together called 'Li and Du'. The 'poetry of Li and Du' was regarded as one of the '100 events that influenced Chinese history' and occupied an important position in Chinese literature and even Chinese history.

Fu Du was an 'Alcohol Sage', stating, 'I am an upright person loving drinking alcohol and have an abhorrence of sin' (*Tour with Ambition* by Fu Du, as depicted in Figure 9.4). 'Bai Li poureth forth a hundred poems after a quart's weight' (*Song on the Eight Faeries in Drinking* by Fu Du) is Fu Du's description of Li Bai and also a true portrayal of himself. 'Feel like a guest when drunk, reciting poems as if getting help from the god' (*Writing Poems While Drinking Alone* by Fu Du). Over 1400 poems by Fu Du are currently available among which over 300 poems are related to alcohol.

The poems of Fu Du are the declarations of his soul and full of his love for the country and concern for people's livelihood. 'The state being broken up, its mounts and streams remain' (*Spring Prospects*) and 'How could there be great hosts of mansions broad to shelter and cheer up scholars all over' (*Song on My Cottage Being Broken by Autumnal Blasts*) are clear evidence. Even the poems about alcohol were concerned about the country and people, stating, 'The mansions burst with alcohols and meat, the poor die frozen on the

Figure 9.4. The hand-drawn story of '*Tour with Ambition*' (The hand drawing is courtesy of Song Zhang, BTBU).

street' (*on the Way from the Capital to Fengxian*) and 'The nobles on the horses are bored of meat and alcohols, yet the poors are deprived of their looms and thatched cottages' (*Tour at the End of the Year*).

Fu Du and Bai Li were friends of poetry as well as alcohol. They once 'slept in one quilt after getting drunk in autumn and play together hand in hand in the day' (*Looking for the residence of Shi Fan with Bai Li*). Fu Du also composed poems about missing Li Bai. For example, 'when could we drink alcohols and discuss poetry again?' (*Memory of Bai Li in spring*).

Fu Du also had drowned his sorrows with alcohol. 'Do not worry about trifles, just drink up all the cups of alcohols in front of you' (*Untitled*). He used to pawn his clothes to buy alcohol. The lines 'When I come back from the imperial court session, I will pawn my clothes to buy alcohols and returned after getting drunk at the river' (*Two Poems of Qujiang*) are similar with those by Bai Li, 'My mottled steed and the fur-lined robe of a thousand crowns. Let my boy lead and fetch out to barter for drinks divine, in order to banish with ye both our griefs eternal trine' (*Carouse, Please*). Fu Du died in a strange land. It was said that his death was the result of drinking, another warning sign.

133. Mu Du and Alcohol

'Upon the Clear and Bright Feast of spring, the rain drizzleth down in spray. Pedestrians on country-side ways in gloom are pining away. When asked 'Where a tavern fair for rest is hereabouts to be found', the shepherd boy the Apricot Bloom Vill doth point to afar and say' (*The Clear-and-Bright Feast* by Mu Du). This is a well-known poem, as described in Figure 9.5, by Mu Du (803–852 A.D.) in the Tang Dynasty. Mu Du is an eminent poet, essayist and calligrapher in the Tang Dynasty, who was called 'Tiny Li and Du' together with Shangyin Li.

Mu Du, the grandson of You Du, the Prime Minister, became a successful candidate in the highest imperial exam. Young and promising, he was the descendant of an eminent family. Mu Du's concern about the country and people was proved by his *Moored on River*

Figure 9.5. The hand-drawn story of '*The Clear-and-Bright Feast*' (The hand drawing is courtesy of Song Zhang, BTBU).

Qinhuai in which he stated, 'Cold water and sand bars veiled in misty moonlight, I moor on River Qinhuai near wineshops at night. The songstress knows not the grief of the captive king, by riverside she sings the song of Parting Spring'.

Good at discussing military matters, Mu Du cared about military affairs and desired to achieve something, which was showed in his *Inscription at Wujiang Pavilion*, 'Win or lose of wars is hardly predictable; a real man takes the insults and stands; Jiangdong has brave talents; a comeback might be made'. However, his official career was not satisfactory. He became an alcoholic who drowned his sorrows with alcohol and lived a life of debauchery. 'Being unrestrained ten years away from home; lingering, lingering to getting cups' (*From Xuancheng to Capital for Inauguration*). 'Waking up from a ten-year dream of Yangzhou; known for fickleness in pleasure quarters' (*Expression of My Heart*).

When we talk about Mu Du, we have to mention another person, Haohao Zhang, a prostitute working in official brothels, who was acquainted with Mu Du in Hongzhou (Nanchang, Jiangxi Province now). They were a perfect match and fell in love at first sight. 'Watching autumn waves in Longsha, touring Zhu Lake in moonlight; often seeing each other, too long even once three days' (*Poem for Zhang Haohao*). Although the talented scholar and the beautiful lady did not have a happy ending, Mu Du still penned down his only scroll of calligraphy *Poem for Haohao Zhang* in her honor, which is collected in the Palace Museum in Beijing.

134. Xiu Ouyang and Alcohol

Xiu Ouyang (1007–1072 A.D.), with a courtesy name of Yongshu and a literary name of Zuiweng (old drunkard), was a famous writer, historian, politician and one of the Eight Great Men of Letters of the Tang and Song Dynasties.

Chuzhou, Anhui Province, is picturesque scenery, 5 km away from the southwest of the ancient city of Chuzhou, Langya Mountain, the historical scenic spot in the east of Anhui and the National Scenic Area, is famous for Zuiweng Pavilion. The highest peak of Langya Mountain rises 317 m above the sea level. Niang Spring drops down between two peaks. Zuiweng Pavilion stays above Niang Spring like a bird spreading the wings. Xiu Ouyang often travelled to Langya Mountain with his friends and colleagues, enjoying alcohol and writing poems in Zuiweng Pavilion, as depicted in Figure 9.6, and even handing official business. *The Story of the Zuiweng Pavilion* was written down and the well-known line 'A drinker minds no cup, but dwells in nature' was accomplished after Xiu Ouyang had consumed alcohol. This line has become a household idiom. Because of Xiu Ouyang, people from Chuzhou are well known today for their capacity for drinking. The line that 'even a sparrow is capable of drinking 150 mL of alcohols' refers to Chuzhou, which reflects the capacity for drinking of the local people.

Xiu Ouyang's life was closely related to alcohol. Many of his poems describe alcohol. A total of 230 poems were included in

Figure 9.6. The hand-drawn story of Xiu Ouyang enjoying alcohol and writing poems with friends in Zuiweng Pavilion (The hand drawing is courtesy of Song Zhang, BTBU).

The Complete Works of Xiu Ouyang, of which the description of alcohol was mentioned 262 times. The following lines reflect the scene of life at the Dragon Boat Festival and the leisurely mood of the author at this festival: 'Pomegranate flowering gorgeous red in May, wetly green poplar hanging down in drizzling. Colorful thread wrapping the polygonal zongzi (a traditional food made from sweet rice), serving on a golden try, to girls of the house with silk covering. Dreaming the Dragon Boat Festival, wearing clean clothes after taking a bath. They raise their cups of realgar wine to ward off poisonous creatures. Orioles singing in tree leaves, breaking stillness and weakening the dreaming girls in Boudoirs.' The following lines expressed the feeling of gathering and separation of friends and loved ones in life: 'Wine cup in hand, I drink to the eastern breeze: Let us enjoy with ease! On the violet pathways green with willows east to the capital. We used to stroll hand in hand in bygone days, Rambling past flower shrubs one and all. In haste to meet and part would ever break the heart. Flowers this year, redder than last appear. Next year more beautiful they'll be. But who will enjoy them with me?' (*Sand-Sifting Waves*).

Xiu Ouyang loved drinking and used to write poems after drinking. He also drowned his sorrows with alcohol, which was demonstrated in his *Writing in Zuiweng Pavilion in Chuzhou*, 'Forty is far from old, the drunkard may write. Drinking empties all minds, no clue of my age/year?...often bringing liquor, approaching ripples from far. Wild birds peeping at my steeping in liquor, stream and cloud persuading me to sleep at the site. Mountain flowers might smell, but can't communicate with me. Only the rock breeze blows me awaking'.

135. Dongpo Su and Alcohol

Shi Su (1037–1101 A.D.), also called Dongpo Su, was a celebrated writer, calligrapher and painter in the Northern Song Dynasty. He made great achievements in poetry, ci, prose, calligraphy and painting. Shi Su, one of the Eight Great Men of Letters of the Tang and Song Dynasties, was called 'Ou-Su' with Xiu Ouyang. The following lines are from *Prelude to Water Melody*, written after he had consumed alcohol: 'How long will the full moon appear? Alcohol cup in hand, I ask the sky. I do not know what time of year it would be tonight in the palace on high. Riding the wind, there I would fly. Yet I am afraid the crystalline palace would be too high and cold for me' and 'Men have sorrow and joy, they part or meet again. The moon is bright or dim and she may wax or wane. There has been nothing perfect since the olden days.' Figure 9.7 depicts the scene of *Prelude to Water Melody*.

Dongpo Su was hospitable and enjoyed treating guests with alcohol. He could not drink much, but had exemplary drinking manners. His poetry written while drinking was full of optimism and expectation of a wonderful life, which was proved by the following lines in *The Beautiful Lady Yu*: 'Holding cups to persuading moon, please being full and no waning. Holding cups to persuading branches, please flowering, blossoming, and no falling, not to leaving the hills.

Figure 9.7. The hand-drawn scene of '*Prelude to Water Melody*' (The hand drawing is courtesy of Song Zhang, BTBU).

Falling in a cup before flowers, no questioning ups and downs. How many people have had this joy? When would one drink if not drinking at flowering?'

Dongpo Su loved drinking, but never indulged in it. He was pleased to raise alcohol cups when treating guests, but disliked drinking day and night, and engaging in social activities. He thought it was tedious work and called it 'hell of drinking and eating'. This was the origin of this idiom.

Dongpo Su was fond of alcohol and made it by himself. He once made honey alcohol and longan alcohol. His *Wine Book of Dongpo,* short but pithy, was a classic book of alcohol-making in China, which included the contents of Qu-making, materials, addition of Qu, addition of materials, fermenting time and alcohol yields.

Dongpo Su also liked to name alcohol and once gave elegant and idyllic names to different alcohols such as 'Wanhuchun', 'Luofuchun', 'Guijiu' and 'Ziluoyijiu'.

When Dongpo Su is mentioned, Dongpo pig knuckle, a traditional Sichuan dish, would have to be mentioned. It is rich in fat but not greasy, soft but not mashed, and with a good color, aroma, taste and appearance. It was said that Fu Wang, Dongpo Su's wife, scorched the pig knuckle carelessly, so she added various condiments to cover the smell of burnt pig knuckle. Surprisingly, the flavor of it improved. Dongpo Su was so happy that he recommended it to his friends. This is the origin of the dish 'Dongpo Pig Knuckle'.

136. Shizhen Li and Alcohol

Shizhen Li (1518–1593 A.D.) is a famous Chinese medicinal and pharmaceutical scientist in the Ming Dynasty. He completed the 1920000-word masterpiece *Compendium of Materia Medica* in nearly 30 years, focusing on the achievements of the Chinese pharmacy before the 16th century. The book is a far-reaching work of medicine and natural history, and is known as the Encyclopedia of China. The book, as shown in Figure 9.8, has been translated and published in English, French, German and Japanese editions.

In *Compendium of Materia Medica*, Shizhen Li elaborated on the history of Chinese alcohol and the efficacy of different alcohols in detail. *Compendium of Materia Medica* pointed out that 'alcohols were made as early as the beginning of the Yellow Emperor, not Di Yi'. This point of view has dated the history of Chinese alcohol about 400 years earlier from the 'Di Yi alcohol-making' time. Although there is still a gap from the 9000-year history now, it was very valuable in the 16th century.

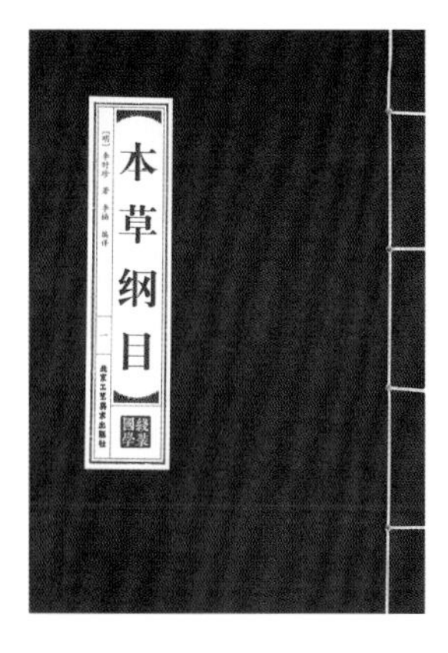

Figure 9.8. The cover of the book, *Compendium of Materia Medica.*

Shizhen Li believed that rice alcohol could be effective against many harmful health risk factors. The aged alcohol could regulate the blood and nourish Qi, warm the stomach to protect against cold and help with respiratory system. The alcohol made in the spring could make people gain weight and could make skin look white.

The exposition of Baijiu (distilled alcohol) from Shizhen Li is also very incisive. He believed that Baijiu could eliminate cold, expel dampness and phlegm, force a person to come out of melancholy, stop diarrhea, treat cholera, malaria and dysphagia, cure cold and pain in the heart and abdomen, alleviate poisoning, kill microorganisms, increase secretion of urine, harden stools, and remove swelling and pain of the eye. However, excessive drinking injures the stomach and liver, and is not conducive to cardiac function, which may reduce life expectancy.

In *Compendium of Materia Medica*, the efficacy and production methods of 69 medicinal alcohols are introduced, including the alcohols made using the raw materials of rehmannia glutinosa, achyranthes bidentata, angelica, wolfberry, ginseng, poria cocos, chrysanthemum, yellow essence, mulberry, ginger, fennel, artemisia, seaweed, bamboo leaf, colorful snake, black snake, viper, turtle meat, tiger bone, deer antler, etc.

Compendium of Materia Medica also revealed that grape wine was good to 'warm waist and kidney, keep the face young looking, and be more resistant to cold'.

The exposition on the positive and negative effects of alcohol from Shizhen Li is objective, which is useful in guiding later generations in healthy drinking habits.

137. Xueqin Cao and Alcohol

When Xueqin Cao (1715–1763 A.D.) is mentioned, his *A Dream of Red Mansions*, the best of the four great masterpieces in Chinese history, has to be mentioned, in which he described many alcohol feasts, introduced a number of alcohol cultures and even vividly depicted drunken characters. His rich knowledge about alcohol comes from his luxurious life when he was young and his affection for alcohol. Xueqin

Cao was born in a servant family of the plain white banner of the Ministry of Internal Affairs in the Qing Dynasty. He lived a rich life in his childhood in Nanjing and moved to Beijing with his family at the age of 13 because the property of his family was confiscated due to a criminal charge. He made a living by selling paintings and calligraphies, and with help from friends. In his old age, his family even ate porridge and bought alcohol on credit. Until his death, alcohol was an indispensable part of his life.

In *A Dream of Red Mansions*, the characters established a poet's club and read poems while enjoying alcohol. In real life, he was talkative and used to write poems while drinking with his friends. The brothers Cheng Guo and Min Guo, two among his poet friends, were famous. *Collection of Sisongtang* by Cheng Guo contained a long poem named *Song of Pawning Sabre for Alcohols*. The preface of it told an interesting story, as depicted in Figure 9.9. Cheng Guo

Note: The Chinese character written on the flag and jar is Baijiu.

Figure 9.9. The hand-drawn story of Cheng Guo exchanging his sabre for alcohol for Xueqin Cao (The hand drawing is courtesy of Song Zhang, BTBU).

encountered Xueqin Cao on his way to visit Min Guo. It was so early that Min Guo had not got up. The two decided to go for a drink, but both of them did not take money. Cheng Guo exchanged his sabre that represented his identity for the alcohol. Xueqin Cao composed a song in gratitude and Cheng Guo responded to it with this long poem. Yiquan Zhang, another friend, also mentioned in his *Poetry Script of Chunliutang* that Xueqin Cao was forthright and liked drinking alcohol. Through those poems, one could learn about Xueqin Cao who was vigorous and unrestrained, and indulged in poetry and alcohol.

Xueqin Cao was addicted to drink rather than simply drowning his sorrows with alcohol. A poem by Min Guo recorded that alcohol helped Cao to paint. Although what he described was Cao's painting of stones, the fiction writing followed the same way. Experiencing the fickleness of human nature, Xueqin Cao expressed his feelings with alcohol, instilled his understanding of the changes of his life into his works and finally finished *A Dream of Red Mansions* with talent and perseverance. After hundreds of years, his book, like a pot of great alcohol, still indulges countless readers.

138. Jin Qiu and Alcohol

Jin Qiu (1875–1907 A.D.), with a courtesy name of Jingxiong and a literary name of Jianhu Female Warrior, was a Bourgeois revolutionist from Shaoxing, Zhejiang not too long ago. She showed special preference for Shaoxing Huangjiu and left many poems and stories about Huangjiu in her short but heroic life. There are 16 poems, 5 ci and several songs about Huangjiu in all her extant works. In her *Song of Sword*, the following lines show an organic unity of poem, alcohol and sword, and the heroic posture and temperament of a female warrior: 'if I were arrested, I would hold sword in my right hand and wine in my left hand. When I get drunk and started to dance, it would be like a dragon or snake walking.' Figure 9.10 depicts the story of *Song of Sword*. The heroic sentiment from the lines in *Carouse, Please* by Li Bai, 'My mottled steed and the fur-lined robe of a thousand crowns. Let my boy lead and fetch out to barter for drinks divine, in order to banish with ye both our griefs eternal trine', can

Figure 9.10. The hand-draw story of '*Song of Sword*' (The hand drawing is courtesy of Song Zhang, BTBU).

also be found in Jin Qiu's *Poem Made While Drinking*, 'I will spend thousands of gold buying a precious blade and trade my mink coat for alcohols. We should cherish our lives and devote to our glorious cause'.

Jin Qiu once wrote down the following lines in her *Japanese Friend Asking for Poem in a Boat in Huanghai and Seeing the Map of Russo-Japanese War*: 'alcohols cannot dispel my tear for my country; the endangered country could be saved when there are many outstanding capable individuals; even it would cost ten thousand lives, our universe has to be saved.' Alcohol with blood and tears and lines full of encouragements made Jin Qiu a tender woman with a strong mind. Poetry and alcohol formed her character even as she was brave with a strong mind. Jin Qiu was a heroine whose fame lives on. Her awe-inspiring uprightness, passion and sacrifice for the country's independence and prosperity will shine forever.

139. Hanzhang Qin, a Great Master of Alcohols

Born in 1908, Hanzhang Qin was a famous expert in the fields of food and fermentation science. Even at the age of 111 years, he loved alcohol drinking, and researched alcohols his entire life and was recognized respectfully as a 'Great Master of Alcohols'. When he was young, he studied at Saint Bruno Agricultural College in Belgium and Berlin University in Germany. He returned to China in 1936 and worked as a professor in Fudan University, Sichuan Provincial College of Education, Central University and Jiangnan University. After the foundation of the People's Republic of China, he worked as an adviser for the Ministry of Food Industry and Ministry of Light Industry, a director for the Research Institute of Food Fermentation Industry of the Ministry of the First Light Industry and the Ministry of Light Industry, a standing member of China Light Industry Association and China Food Industry Association, and a deputy to the 3rd, 5th and 6th National People's Congress. He was also the honorary president of the Baijiu Association of China Food Industry Association and received a special government subsidy of the State Council.

Hanzhang Qin directed the experimental work of Fen Baijiu, which was a milestone for the Baijiu industry in the 1960s. At that time, the Chinese economy was in ruins and was ready for a revival. It was Qin, with other researchers, who comprehensively investigated every aspect in Fen Baijiu production with scientific methodologies, established a group of chemical testing methods and identified some important microbial strains that greatly improved the production and quality of the alcoholic products. This was the first time that researchers theoretically and systematically studied and summarized the experiences of Baijiu production handed down from the ancestors through the ages. It was also the first time that applied science and technology had improved the traditional technologies. In addition, Mr. Qin made a great contribution to the improvement and extension of the production techniques for Huangjiu and beer.

After his retirement at the age of 82, Mr. Qin still took an active part in seminars and conference activities related to the alcohol industry. Besides, he practiced calligraphy and wrote poems, and created many poems with 7 characters in a line about the alcohol culture and a healthy life that developed and expanded the traditional alcohol culture. In the meantime, he spent more time on writing and transferred what he had learnt and thought into words for later generations. According to the statistics, he finished over 40 monographs about the manufacturing technique of Baijiu, the scientific technology of Baijiu and alcohol culture, which have greatly advanced the development of the alcohol-making industry in China.

Bibliography

Administration for Marked Regulation of Guizhou Province. DB52/T 550-2013 Dong-flavor type Baijiu (in Chinese).

Awareness Qin. Xueqin Cao and Liquor. *A Dream of Red Mansions*, 1991, (01): 42–43 (in Chinese).

Chai, C. Langyatai Baijiu, the first brand of Chinese 'he' culture. *Business Weekly*, 2014, (14): 62–63 (in Chinese).

Chen, B. History of Yubing Shao Baijiu in Shiwan distillery. *Liquor Making*, 1984, (02): 65–57 (in Chinese).

Chen, L., Yang, C., Huang, W., *et al.* Advances on applications and development of the steaming bucket distillation in liquor production. *The Food Industry*, 2016, 37(08): 222–225 (in Chinese).

Chen, T. On the characteristics and inheritance of aged fragrant from Fen and Huanghelou Baijiu. *Chian Wine News*, 2017-6-27 (A15) (in Chinese).

Chen, X., and Wang, Y. Characteristic techniques & environmental factors of "Yanghe Blue Classic" liquor & the relations of its microconstituents and people's health. *Liquor-Making Science & Technology*, 2007, (8): 161–164 (in Chinese).

Cheng, G. Research on brand innovation strategy of Xinghuacun Fen Baijiu, the China time honored brand. *Beijing: Capital University of Economics and Business*, 2011 (in Chinese).

Chu, J. The culture and literature writing of Baofeng Baijiu. *Science and Technology*, 2016, 26(28): 255–257 (in Chinese).

Cui, L. Biosynthetic Pathways and Steps of Pyrazine Compounds in Maotai-flavor Liquor. *Liquor Making*, 2007, (05): 39–40 (in Chinese).

Cui, L. Nutrition component of Chinese liquor and its benefit to human health. *Liquor Making*, 2008, 35, 15-18 (in Chinese).

Dai, J., Chen, S., Xie, G., *et al.* Isolation and sequence analysis of angiotensin converting enzyme inhibitory peptides in chinese rice wine. *Journal of Instrumental Analysis*, 2006, (04): 74–77 (in Chinese).

Dong, S. Ancient Chinese alcohols culture. *Beijing: China Bookstore press*, 2012: 178–179 (in Chinese).

Editorial Committee of Encyclopedia of China. Encyclopedia of China. Beijing: Encyclopedia of China Publishing House, 2002: 840 (in Chinese).

Fan, H., Fan, W., and Xu, Y. Characterization of key odorants in Chinese chixiang aroma-type liquor by gas chromatography-olfactometry, quantitative measurements, aroma recombination, and omission studies. *Journal of Agricultural and Food Chemistry*, 2015, 63: 3660–3668.

Fan, W., and Qian, M. Characterization of aroma compounds of Chinese "Wuliangye" and "Jiannanchun" liquors by aroma extract dilution analysis. *Journal of Agricultural and Food Chemistry*, 2006, 54(7): 2695-2704.

Fan, W., and Xu, Y. Identification of volatile components of Fenjiu and Langjiu by Liquid-Liquid Extraction and normal phase chromatography (Part 2). *Liquor-Making Science & Technology*, 2013, (3):17–27 (in Chinese).

Fan, W., and Xu, Y. Review of functional factors and quality safety factors of baijiu (Chinese Liquor). *Liquor-Making Science & Technology*, 2012, (3): 17–22 (in Chinese).

Fan, W., and Xu, Y. Scientifically understand the biological active ingredients in Chinese liquor. *Liquor-Making Science & Technology*, 2013, (9): 1–6 (in Chinese).

Fan, W., Hu, G., and Xu, Y. Analysis of aroma components in Chinese herbaceous aroma type liquor. Journal of Food Science and Biotechnology, 2012, 31(08): 810–819 (in Chinese).

Fan, W., Xu, Y., Yang, T., *et al.* Volatile compounds of supple and mellow flavor type in yanghe's lansejidian liquors detected by liquid-liquid extraction coupled with fractionation. *Liquor Making*, 2012, (1): 21–29 (in Chinese).

Fu, G. Realistic and innovative research on xifeng liquor making technology. *Niang Jiu*, 2016, 43(04): 9–14 (in Chinese).

General Administration of Quality Supervision, Inspection and Quarantine of the People's Republic of China. GB/T 10781.2-2006 Mild-flavour Chinese Spirits (in Chinese).

General Administration of Quality Supervision, Inspection and Quarantine of the People's Republic of China. GB/T 10781.3-2006 Rice-flavour Chinese Spirits (in Chinese).

General Administration of Quality Supervision, Inspection and Quarantine of the People's Republic of China. GB/T 14867-2007 Feng-flavour Chinese Spirits (in Chinese).

General Administration of Quality Supervision, Inspection and Quarantine of the People's Republic of China. GB/T 16289-2018 Chi Xiang Xing Baijiu (in Chinese).

General Administration of Quality Supervision, Inspection and Quarantine of the People's Republic of China. GB/T 20823-2017 Te Xiang Xing Baijiu (in Chinese).

General Administration of Quality Supervision, Inspection and Quarantine of the People's Republic of China. GB/T 20824-2007 Zhima-flavour Chinese Spirits (in Chinese).

General Administration of Quality Supervision, Inspection and Quarantine of the People's Republic of China. GB/T 20825-2007 Laobaigan-flavour Chinese Spirits (in Chinese).

General Administration of Quality Supervision, Inspection and Quarantine of the People's Republic of China. GB/T 22736-2008 Product of geographical indication-Jiugui liquor (in Chinese).

General Administration of Quality Supervision, Inspection and Quarantine of the People's Republic of China. GB/T 23547-2009 Nong Jiang-flavour Chinese Spirits (in Chinese).

General Administration of Quality Supervision, Inspection and Quarantine of the People's Republic of China. GB/T 26760-2011 Jiang-flavour Chinese Spirits (in Chinese).

General Administration of Quality Supervision, Inspection and Quarantine of the People's Republic of China. GB/T 10781.1-2006 Strong-flavour Chinese Spirits (in Chinese).

Gou, M., Wang, H., Yuan, H., *et al.* Characterization of the microbial community in three types of fermentation starters used for Chinese liquor production. *Journal of the Institute of Brewing*, 2016, 37, 94–98.

Guang, J., and Gao, L. Study on the production techniques of luzhou-flavor yingjia gongjiu. *Liquor-Making Science & Technology*, 2009, (04): 76–78 (in Chinese).

Guo, G., He, M., Zou, J., *et al.* Extraction and Isolation of Tartary Buckwheat Flavonoids and its Antioxidant Activity. *Journal of Food Science and Technology*, 2008, 29(12): 373–376 (in Chinese).

Guo, X. Tartary Buckwheat Polyphenols and its Improvement in Endothelial Insulin Resistance. Yang Ling: Northwest A & F University, 2013.

Guo, X. The research of alcohol industry development and social cultural changes in modern china. Wu Xi: Jiang Nan University, 2015 (in Chinese).

Han, F., and Xu, Y. Identification of low molecular weight peptides in chinese rice wine (Huang Jiu) by UPLC-ESI-MS/MS. *Journal of the Institute of Brewing*, 2011, 117(2): 238–250.

Han, Q., Shi, J., Zhu, J., *et al.* Enzymes extracted from apple peels have activity in reducing higher alcohols in Chinese liquors. *Journal of Agricultural and Food Chemistry*, 2014, 62, 9529–9538.

Happy news: Xifeng Baijiu is regarded as a cultural relic[EB/OL]. 2017-09-019 [2018-09-15]. http://spirit.tjkx.com/detail/1042667.html (in Chinese).

He, Z. To View All the Flowers of Chang'an in One Day: Tang Poems in Original Phyme. Beijing: CITIC Press, 2017.

Hou, G. Study on the construction of the brand of the Liquor New Lang. Cheng Du: University of Electronic Science and Technology of China, 2015 (in Chinese).

Hu, J., Cai, G., and Liu, Y. Investigation on Jiuhai (Liquor Storage Container Weaved by Twigs of the Chaste Tree). *Liquor-Making Science & Technology*, 2008, (09): 118–119 (in Chinese).

Hu, Y., and Xu, X. Research progress of ferulic acid in chemistry and pharmacology. *Chinese Patent Medicine*, 2006, 28(2): 253–255.

Huang, P. Leading scholar in chinese wine industry — QIN Han-zhang. *Liquor-Making Science & Technology*, 2007, (04): 19–26 (in Chinese).

Huang, P., Jiang, Y., Zhang, X., *et al.* Celebration for Mr. QIN Hanzhang's 109th Birthday. *Liquor-Making Science & Technology*, 2017, (02): 17–22 (in Chinese).

Huang, W. Extraction, spearation, compsition and antioxidant activity of polysaccharides from hakka rice wine. Guang Zhou: Zhongkai University of Agriculture and Engineering, 2017 (in Chinese).

Javier, B. Solid-state fermentation: Physiology of solid medium, its molecular basis and applications. *Process Biochemistry*, 2012, 47, 175–185.

Ji, L. Product of geographical indication: Niulanshan Erguotou Baijiu. *China Standardization*, 2008, (08): 72–73 (in Chinese).

Jia, L., and Wei, L. Pharmacological characteristics and clinical application of lovastatin. *Chinese Journal of Clinical Rational Drug Use*, 2015, 8(15): 14–15 (in Chinese).

Jia, Q. A report on the establishment of “Dong-flavour” liquor. *Liquor-Making Science & Technology*, 1999, (05): 75–79 (in Chinese).

Jiang, F. Study on Extraction and Activity of Active Components in Lees . Changchun: Jilin University, 2009 (in Chinese).

Jiang, X., and Chai, C. Langyatai Baijiu: ‘difference ahead’. *Business Weekly*, 2010, (05): 64–66 (in Chinese).

Jin, G., Zhu, Y., and Xu, Y. Mystery behind Chinese liquor fermentation. *Trends in Food Science & Technology*, 2017, 63, 18–28.

Jin, Y., and Xu, H. Suffering and glory: Where does the power of the Communist Party of China come from?. Fuzhou: The Straits Publishing, 2013: 230–232 (in Chinese).

Ke, Y. Study on the Aromatic Compounds in Kinmen Sorghum Liquor. Xia Men: Jimei University, 2016 (in Chinese).

Lai, A., Zhao, D., and Cao, J. History, status and development trend of zhima-flavor chinese Spirits. *Liquor Making*, 2009, 36(01): 91–93 (in Chinese).

Li, A., Xu, X., Tang, Y., *et al.* Research on health functional components of Gujing Gongjiu based on full two-dimensional gas chromatography-time-of-flight mass spectrometry analysis. *Liquor-Making Science & Technology*, 2016, (1): 50–52 (in Chinese).

Li, C. Jingzhiguniang Baijiu drunk world long. *China Drinks*, 2002, (04): 56–57 (in Chinese).

Li, C., Zhang, H., Zhao X., *et al.* Analysis of Wuling wine production process innovation. *Liquor-Making Science & Technology*, 2009 (in Chinese).

Li, D. Relationship between koji, cellar, process and production and quality of luzhou-flavor liquor. *Liquor Making*, 2008, (04): 3–9 (in Chinese).

Li, D. Training course of Baijiu brewing (for distiller, baijiumaker and taster). Beijing: China Light Industry Press, 2013 (in Chinese).

Li, F. Research on the breakthrough growth of Baijiu sales in Sichuan tuo-paiqu Liquor Co., Ltd. Chengdu: Southwestern University of Finance and Economics, 2009 (in Chinese).

Li, G., and Li, D. Historical changes of shanxi brewing industry. *Journal of Northwest University: Nature Scientific Edition*, 2010, 40 (05): 929–933 (in Chinese).

Li, H., Hu, X., Li, A. *et al.* Headspace solid phase microextraction and stir bar adsorption Analysis of Aroma Components in Gujing Distillery by

Extraction Technology. *Food Science*, 2017, 38(04): 155–164 (in Chinese).

Li, H., Liu, J., Liang, J., *et al.* Study on volatile components in 2 Gujinggong Liquors. *Journal of Food Science and Technology*, 2016, 34(1): 55–65 (in Chinese).

Li, S. Compendium of Materia Medica (Volume II). Beijing: Huaxia Press, 2002: 1045–1052 (in Chinese).

Li, S. Discussion on the characteristics of songhe grain and liquid style. *Liquor Making*, 2008, (02): 30–34 (in Chinese).

Li, X. Discussion on songhe liquor production process and its style characteristics. *Liquor Making*, 2014, 41(04): 70–74 (in Chinese).

Li, Y. Niulanshan Erguotou: Leaving a strong color for Chinese liquor culture. China Business News, 2012-10-19(016) (in Chinese).

Li, Y. Study on the fingerprints of the flavor substances of Fuyuxiang Jiugui Liquor. Changsha: Hunan University, 2011 (in Chinese).

Liang, H. The story behind the wine and the World Expo. *Shanxi Archives*, 2010, (03): 51–54 (in Chinese).

Liao, S., Kao, T., Chen, W., *et al.* Tetramethylpyrazine reduces ischemic brain injury in rats. *Neuroscience Letters*, 2004, 372: 40–45.

Liu, C., Lin, C., Ng, L., *et al.* Protection by tetramethylpyrazine in acute absolute ethanol-induced gastric lesions. *Journal of Biomedical Science*, 2002, 9: 395–400.

Liu, D., Zhao, Z., Yi, R., *et al.* Application of wine in medicine. *Chinese Wine*, 2002, (5): 32–33 (in Chinese).

Liu, H., and Sun, B. Effect of fermentation processing on the flavor of baijiu. *Journal of Agricultural and Food Chemistry*, 2018, 66(22): 5425–5432.

Liu, J. Research on the construction of four special wine brands. Nanchang: Nanchang University, 2013, 231 (in Chinese).

Liu, M. Study on the development of liquor industry cluster in Hubei Province from the perspective of scientific development. Wuhan: Wuhan Institute of Technology, 2011 (in Chinese).

Liu, M., Tang, Y., Guo, X., *et al.* Deep sequencing reveals high bacterial diversity and phylogenetic novelty in pit mud from Luzhou Laojiao cellars for Chinese strong-flavor Baijiu. *Food Research International*, 2017, 102, 68–76.

Liu, M., Tang, Y., Zhao, K., *et al.* Determination of the fungal community of pit mud in fermentation cellars for Chinese strongflavor liquor, using DGGE and Illumina MiSeq sequencing. *Food Research International*, 2017, 91, 80–87.

Liu, M., Wang, J., Sun, P., *et al.* Application of bran in the production of sesame-flavor liquor. *Liquor-Making Science & Technology*, 2013, (03): 69–70, 74 (in Chinese).

Liu, X., Liu, H., Zeng, R., *et al.* Research progress on the aroma characteristics of Dong wine flavor. *Liquor-Making Science & Technology*, 2016, (12): 91–93 (in Chinese).

Lu, Y., Li, Y., Huang, J., *et al.* Classification, production process and nutritional value of chinese liquor. *Agricultural Engineering Technology (Agricultural Product Processing)*, 2007, 1, 21–24.

Luo, T., Fan, W., and Xu, Y. Characterization of volatile and semi-volatile compounds in Chinese rice wines by headspace solid phase microextraction followed by gas chromatography-mass spectrometry. *Journal of the Institute of Brewing*, 2008, 114(2): 172–179.

Luo, Z., Zeng, M., Chen, D., *et al.* The excavation briefing of the Mianzhu Jiannanchun Winery Site in 2004. *Sichuan Cultural Relics*, 2007, (02): 3–12, 97–98 (in Chinese).

McGovern, P., Zhang, J., Tang, J., *et al.* Fermented Beverages of Pre- and Proto-historic China. *Proceedings of the National Academy of Sciences of the United States of America*, 2004, 101(51): 17593–17598.

Mo, X., Xu, Y., and Fan, W. 4-vinyl guaiacol and vanilla during storage Changes in aldehydes and influencing factors. *Food and Fermentation Tech.* 2016, 42(02): 29–34 (in Chinese).

Ou, S., Bao, H., and Lan, Z. Research progress of pharmacological effects of ferulic acid and its derivatives. *Chinese Herbal Medicine*, 2001, 24(3): 220–221.

Ou, S. Function and application of ferulic acid. *Modern Food Technology*, 2002, 18(4): 50–53 (in Chinese).

Peng, J., Mao, J., Huang, G., *et al.* Anti-oxidation activities of rice wine polysaccharides in vitro. *Science and Technology of Food Industry*, 2012, 33(20): 94–97 (in Chinese).

Shen, C. The separation and extraction, biological activities of polysaccharides from Shaoxing rice wine and their effects on intestinal microflora. Wuxi: Jiangnan University, 2014 (in Chinese).

Shen, M., Zhang, C., and Wang, Y. Research progress on microbial microbiology of liquor. *China Brewing*, 2016, 35(05): 1–5 (in Chinese).

Shen, Y. Baiqing Production Technology Book. Beijing: China Light Industry Press, 2015 (in Chinese).

Shen, Y. Liquor Production Technology Book. Beijing: China Light Industry Press, 2013 (in Chinese).

Shen, Y. The distillation of liquor in the barrel. *Liquor Making*, 1995, (05): 7–18 (in Chinese).

Shen, Z. Shuze mengde cao. *China Drinks*, 2017, (08): 58 (in Chinese).

Shi, Z. The Long March. Beijing: The Communist Party History Press, 2016: 126–127 (in Chinese).

Su, L. The wine complex in ouyang Xiu's Poetry. *Journal of Chuzhou University*, 2012, (06): 1–5 (in Chinese).

Sun, B., Li, H., Hu, X., *et al.* The development trend of healthy Baijiu. *Journal of Chinese Institute of Food Science and Technology*, 2016, 8, 1-8 (in Chinese).

Sun, B., Wu, J., Huang, M., *et al.* Research progress of liquor flavor chemistry. *Journal of Chinese Institute of Food Science and Technology*, 2015, 15(9): 1–8 (in Chinese).

Sun, D. An Anthology of Ancient Chinese Poetry and Prose. Shanghai: Shanghai Joint Publishing Press, 2019.

Sun, J., and Ma, L. Marley History and Wine Culture Research. Beijing: Social Sciences Academic Press, 2012: 8–9, 54–154, 180–184 (in Chinese).

Sun, T. The 100 events affecting Chinese history. Beijing: Line Pack Book Office, 2003: 184–189 (in Chinese).

Sun, X., Wang, X., Liu, M., *et al.* Determination of tetramethylpyrazine, 4-methylguaiacol and 4-ethylguaiacol in 67 kinds of liquors by vortex-assisted liquid-liquid microextraction combined with GC-MS. *Food Technolgy*, 2017, (18): 73–79 (in Chinese).

Sun, X., Zhang, F., Dong, W., *et al.* GC-FPD analysis of 3-methylthiopropanol in sesame-flavor liquor. *Journal of Food Science and Technology*, 2014, 32(5): 27–34 (in Chinese).

Travel notes of Baijiu. Magnificent Jingzhi likes a poem-tell you about travel notes of Jingzhi in my eyes[EB/OL]. 2017-09-07 [2018-09-15]. http:// www.jianiang/cn/jiuwebhua/jiulv/0zm91512017. html. (in Chinese)

Wang, B., and Wang, H. Study on the technology of producing bran and sauce-flavored liquor with soy sauce. *Liquor Making*, 2016, 43(03): 89–93 (in Chinese).

Wang, B., Li, H., Zhang, F., *et al.* Analysis of nitrogen-containing compounds of Guojing Sesame-flavour liquor by liquid-liquid extraction coupled with GC-MS and GC-NPD. *Food Science*, 2014, 35(10): 126–131 (in Chinese).

Wang, B., Xin, C., Han, J., *et al.* Analysis of nitrogen compounds in sesame-flavor liquor by headspace solid-phase microextraction combined with GC/NPD technology. *Journal of Chinese Institute of Food Science and Technology*, 2015, 15(4): 247–253 (in Chinese).

Wang, C., Dong, J., and Guo, L. Microorganisms in Daqu: A starter culture of Chinese maotai-flavor liquor. *World Journal of Microbiology & Biotechnology*, 2008, 24(10), 2183–2190.

Wang, C., Shi, D., and Gong, G. Microorganisms in Daqu: A starter culture of Chinese maotai-flavor liquor. *World Journal of Microbiology & Biotechnology*, 2008, 24, 2183–2190.

Wang, F. On Ruan Ji's and Cao Xue qing's "drunken Life". *Journal of Xinyu College*, 2009, 14 (03): 49–51 (in Chinese).

Wang, J. Gujing Distillery: A Historic Wine. *Decision-making Exploration*, 2011, (12): 91 (in Chinese).

Wang, J. Study on the productive protection of traditional brewing technology of Shuijingfang Baijiu. Chengdu: Sichuan Academy of Social Sciences, 2014 (in Chinese).

Wang, J., Liu, L., Ball, T., *et al.* Revealing A 5,000-y-old beer recipe in China. *Proceedings of the National Academy of Sciences of the United States of America*, 2016, 113(23): 6444–6448.

Wang, K. Inheritance & Development of Time-honored Brand Shuang' gou Liquor. *Liquor-Making Science & Technology*, 2011, 05: 125–126 (in Chinese).

Wang, R. Enjoy the Baijiu capital and find the ancient beauty in Yanghe River. The People's Daily, 2015-09-25(015) (in Chinese).

Wang, S., Wang, Q., Lu, L., *et al.* Research progress of the microbial diversity, enzyme system and formation of flavor compounds in chinese flavor liquor. *Journal of Agricultural Biotechnology*, 2017, 25(12): 2038–2051 (in Chinese).

Wang, X. Collection, arrangement and publication of local cultural documents in the library. *Library World*, 2015, (04): 70–72 (in Chinese).

Wang, X., Fan, W., and Xu, Y. Comparison on Aroma Compounds in Chinese Soy Sauce and Strong Aroma type Liquors by Gas Chromatography-Olfactometry, Chemical Quantitative and Odor Activity Values Analysis. *European Food Research and Technology*, 2014, 239(5): 813–825.

Wang, Y., Luo, H., and Wang, C. Research progress in microbes for the production of xiaoqu. *Liquor-Making Science & Technology*, 2014, (04): 78–82 (in Chinese).

Wang, Y., Zhang, C., Li, H., *et al.* Multiple microbial systems of composite functional starter for soft-type Baijiu(Liquor) production. *Liquor-Making Science & Technology*, 2015, (04): 41–45 (in Chinese).

Wang, Z., Zhang, S., Zhao, J., *et al.* Headspace solid-phase micro extraction-gas chromatography-mass spectrometry analysis of volatile components of bamboo leaf green wine. *Journal of Food Science and Technology*, 2014, 35(8): 253–258 (in Chinese).

Wei, L. The way of good Baijiu: The contract between heaven and man. China Wine News, 2016-06-07 (A11) (in Chinese).

Wu, G. Design of Jingzhi Wine Annals Books and Their Derivatives. Shandong: Shandong University of Art and Design, Jinan, 2017 (in Chinese).

Wu, J., and Xu, Y. Comparison of pyrazine compounds in seven Chinese liquors using headspace solid-phase micro-extraction and GC-nitrogen phosphourus detection. *Food Science & Biotechnology*, 2013, 22(5): 1–6.

Wu, J. Study on functional composition in chinese liquor- tetramethylpyrazine. *Liquor Making*, 2006, (06): 13–16 (in Chinese).

Wu, J., Huang, M., Sun, B., *et al.* Analysis of volatile compounds in Jingzhi Baigan Liquor by Liquid-liquid Extraction (LLE) and Gas Chromatography-Mass Spectrometry (GC-MS). *Journal of Chinese Institute of Food Science and Technology*, 2014, 35(8): 72–75 (in Chinese).

Wu, J., Huo, J., Huang, M., *et al.* Structural Characterization of a Tetrapeptide from Sesame Flavor-Type Baijiu and its Preventive Effects Against AAPH-Induced Oxidative Stress in HepG2 Cells. *Journal of Agricultural and Food Chemistry*, 2017, 65(48): 10495–10504.

Wu, J., Sun, B., Luo, X., *et al.* Cytoprotective Effects of a Tripeptide from Chinese Baijiu against AAPH-Induced Oxidative Stress in HepG2 Cells Via Nrf2 Signaling. *RSC Advances*, 2018, 8: 10898–10906.

Wu, J., Sun, B., Zhao, M., *et al.* Discovery of a bioactive peptide, an angiotensin converting enzyme inhibitor in chinese baijiu. *Journal of Chinese Institute of Food Science and Technology*, 2016, 16(09): 14–20 (in Chinese).

Xi, B., Veeranki, S., Zhao, M., *et al.* Relationship of alcohol consumption to All-cause, cardiovascular, and cancer- related mortality in U.S. adults. *Journal of the American College of Cardiology*, 2017, 70(6): 913–922.

Xia, Y. Guilin Sanhua Baijiu. *Food and Fermentation Industries*, 1980, (04): 40 (in Chinese).

Xie, M., and Wu, Y. Inheritance and innovation of the production techniques of soybean-flavor liquor. *Liquor-Making Science & Technology*, 2012, (08): 82–83 (in Chinese).

Xin, Z., and Chen, J. Chinese Baijiu culture. Jinan: Shandong Education Press, 2009 (in Chinese).

Xing, M. Analysis of baofeng baijiu and light flavor type baijiu. *Liquor Making*, 1987, (06): 17–19 (in Chinese).

Xiong, X. Summary of the production techniques of maotai-luzhou-flavor baiyunbian liquor. *Liquor-Making Science & Technology*, 2007, (09): 35–42 (in Chinese).

Xiong, Y. A Study on Brand Development Strategy of Luzhou Laojiao of Master Thesis. Sichuan: Chengdu University of Electronic Science and Technology of China, 2016 (in Chinese).

Xiong, Z. Development of Te-type Liquor--Record on the study of the approaches to Improve Si'te liquor quality. *Liquor-Making Science & Technology*, 2006, (01): 102–104 (in Chinese).

Xu, J. Developing a Rice Wine with high content of Monacolin K. NanJing: Nanjing Agricultural University, 2004 (in Chinese).

Xu, Y. 300 Song Lyrics. Beijing: China Intercontinental Press, 2018.

Xu, Y. 300 Tang Poems. Beijing: China Intercontinental Press, 2018.

Xu, Y. Book of Poetry. Beijing: China Intercontinental Press, 2011.

Xu, Y. Decoding of liquor culture in Yibin history archives. *Liquor-Making Science & Technology*, 2016, (08): 126–128 (in Chinese).

Xu, Y. Tang Poetry in Paintings. Beijing: China Translation & Publishing House, 2017.

Xu, Y., Zhang, R., Wu, Q., *et al.* Identification and function study of lipopeptide compounds, A biologically active substance in liquor. *Liquor-Making Science & Technology*, 2014, (12): 1–4, 7 (in Chinese).

Xu, Z. The unique flavor features of jian' nanchun liquor & its classic production techniques. *Liquor-Making Science & Technology*, 2010, (11): 53–56 (in Chinese).

Xu, Z., Chen, Y., Zhou, Z., *et al.* Research on healthy functional ingredients in Chinese famous wine jiannanchun. *Liquor-Making Science & Technology*, 2008, (5): 41–44 (in Chinese).

Yang, D., Jiang, Y., and Deng, W. Technical improvement & innovation and quality control of safflower lang liquor. *Liquor-Making Science & Technology*, 2007, (03): 54–57 (in Chinese).

Yang, H., Yang, L., Chai, Y., *et al.* Comparison of antioxidant activity and the content of free and bound phenolics in common buckwheat and tartary buckwheat particles. *Science and Technology of Food Industry*, 2011, 32(5): 90–94 (in Chinese).

Yang, T., Li, G., Wu, L., *et al.* Research on Chinese baijiu health factors and their breeding strain selection and application in production (I) Research on Chinese Liquor Health Factors. *Liquor-Making Science & Technology*, 2010, (12): 65–69 (in Chinese).

Yang, Y. The Research of Brand Strategy for Kinmen Kaoliang Liquor in White Liquor Market of Mainland China. Xiamen: Xiamen University, 2009 (in Chinese).

Yang, Z., and Gong, X. Being the guider of chinese cultural liquor-the power displayed by jiugui liquor cultural management strategy. *Liquor-Making Science & Technology*, 2002, (01): 91–94 (in Chinese).

Yu, P. Setting-up of HACCP system group co ltd. of Henan Baofeng wine industry. Jinan: Shandong University, 2005 (in Chinese).

Yu, Q. Traditional Baijiu brewing technology. Beijing: China Light Industry Press, 2013: 3–6 (in Chinese).

Yu, Q. Traditional Baijiu brewing technology. Beijing: China Light Industry Press, 2015 (in Chinese).

Zeng, J., Luo, Z., and Zhang, J. Ligustrazine in treating 68 cases of diabetic nephropathy. *Guangdong Medical Journal*, 2005, 26(7): 1004–1005 (in Chinese).

Zhai, W. Chinese alcohols ceremony. Shanghai: Shanghai Popular Science Press, 2011 (in Chinese).

Zhang, C. Investigation of off-Flavor Compoundsin Chinese Liquor. Wu Xi: Jiangnan University, 2013 (in Chinese).

Zhang, C. The Relationship between Quality, Microorganisms and Flavour Components of Luzhou Laojiao Daqu. Wu Xi: Jiangnan University, 2012 (in Chinese).

Zhang, D., and Wang, H. 100 influential people in Chinese history. Beijing: The Ethnic Publishing Press, 1999: 170–172 (in Chinese).

Zhang, E. Research on the characteristic aroma components of Luzhou-flavored Yanghe Sky Blue and Qingluo Erguotou Daqu. Wuxi: Jiangnan University, 2009 (in Chinese).

Zhang, F. The Caproic Acid Bacteria and Physical and Chemical Indicators of Shuanggou Luzhou-Flavor Liquor Difference in the New and Old Pits. NanJing: Nanjing Agricultural University, 2013 (in Chinese).

Zhang, J. Research on Brand Marketing Strategies of S Wine Company. Dalian: Dalian University of Technology, 2013 (in Chinese).

Zhang, J. The Analysis on the Wine Culture of the Wei Jin and Southern and Northern Dynasties Period. Jinan: Shandong Normal University, 2010 (in Chinese).

Zhang, J., Cui, C., Tong, Z., *et al.* Baijiu production process and technology. Beijing: Chemical Industry Press, 2014: 106–118, 148–168 (in Chinese).

Zhang, L., and Shen, C. The Brewing Technique Pandect of Lu-Type Liquor. Beijing: China Light Industry Press, 2011: 464–482 (in Chinese).

Zhang, M. Analysis of the success of Jingzhi Liquor industry Co.Ltd. from the angle of culture. *Liquor-Making Science & Technology*, 2013, 08: 106–108 (in Chinese).

Zhang, R., Liu, R., Chen, R., *et al.* Isolation and identification of substances with angiotensin converting enayme inhibitory activity in fujian rice wine. *Journal of Fuzhou University (Natural Science Edition)*, 1996, (06): 114–118 (in Chinese).

Zhang, S., and Xu, D. The Brewing Technique Pandect of Lu-Type Liquor. Beijing: China Light Industry Press, 2011: 2–7 (in Chinese).

Zhang, S., Liu, Y., Zhu, B., *et al.* Determination of polysaccharide content in distillers grains by a-naphthol-sulfuric acid method. *Journal of Food Science and Technology*, 2013, 34(18): 245–248 (in Chinese).

Zhang, W. Research on the development of Quanxing Baijiu. Chengdu: Southwestern University of Finance and Economics, 2003 (in Chinese).

Zhang, Y. Review of the development process of full-bodied-flavor liquor. *Liquor-Making Science & Technology*, 2011, 10: 117–121 (in Chinese).

Zhang, Z., Fan, W., and Xu, Y. Comparative analysis of free Amino Acids in different flavor liquors. *Science and Technology of Food Industry*, 2014, 35(17): 280–284 (in Chinese).

Zhao, D., Li, Y., and Xiang, S. Determination of aromatic aroma components in lees and liquor by gas chromatography-mass spectrometry. *Liquor-Making Science & Technology*, 2006, (10): 92–94 (in Chinese).

Zhao, F. Study on the sustainable development of Chinese Baijiu industry. Beijing: Chinese Academy of Social Sciences, 2014 (in Chinese).

Zhao, J., Han, X., Yang, H., *et al.* Preliminary research on fermentation mechanization of fen-flavor's ground-pot. *Food and Fermentation Industries*, 2013, 39(11): 81–84 (in Chinese).

Zheng, X., and Han, B. Baijiu, Chinese liquor: History, classification and manufacture. *Journal of Ethnic Foods*, 2016, 3, 19–25.

Zheng, X., Yan, Z., Robert Nout, M., *et al.* Microbiota dynamics related to environmental conditions during the fermentative production of Fen-Daqu, a Chinese industrial fermentation starter. *International Journal of Food Microbiology*, 2014, 182–183, 57–62.

Zheng, Y., Sun, B., Zhao, M., *et al.* Characterization of the Key Odorants in Chinese Zhima Aroma-type Baijiu by Gas Chromatography-olfactometry, Quantitative Measurements. Aroma Recombination and Omission Studies. *Journal of Agricultural and Food Chemistry*, 2016, 64(26): 5367–5374.

Zhong, G., Zou, H., and Zhou, R. Discussion on the modernization and internationalization of chinese liquor. *Liquor-Making Science & Technology*, 2012, (1): 82 (in Chinese).

Zhong, Y., Cui, R., and Teng, K. Investigation on the Formation of "Soft" Quality of Yanghe Blue Classic Liquor (Part I). *Liquor-Making Science & Technology*, 2009, 04: 117–121, 126 (in Chinese).

Zhong, Y., Cui, R., and Teng, K. Investigation on the Formation of "Soft" Quality of Yanghe Blue Classic Liquor (Part II). *Liquor-Making Science & Technology*, 2009, 05: 121, 126 (in Chinese).

Zhou, S. Famous Baijiu debate: From Maotai town to Xinghua Village. China Wine News, 2015-03-03 (B31) (in Chinese).

Zhou, X. Huanghelou: Brew the perfect state of 'harmony between man and nature'. China Wine News, 2013-10-15 (A23) (in Chinese).

Zhou, X., Chen, X., Li, L., *et al.* Study on the relationship between the flavor characteristics of yanghe soft liquor and human health. *Liquor-Making Science & Technology*, 2014, (11): 31–34 (in Chinese).

Zhou, X., Yang, Z., Liu, X., *et al.* Rapid analysis of the content of main components of lees by near infrared reflectance spectroscopy. *Journal of Agricultural Machinery*, 2012, 43(3): 103–107.

Zhu, H. Jiannanchun distillery site selected as one of the 'top ten archaeological discoveries in China'. Science and Technology Daily, 2005-06-06 (in Chinese).

Zhu, S., Lu, X., Ji, K., *et al.* Characterization of flavor compounds in Chinese liquor Moutai by comprehensive two-dimensional gas chromatography/time-of-flight mass spectrometry. *Analytica Chimica Acta*, 2007, 597, 340–348.

Zhu, Z. Shuang'gou Daqu Quite Popular During the Period of Republic of China. *Liquor-Making Science & Technology*, 2012, (02): 116–117 (in Chinese).

Zhuang, M., and Chen, H. Multi-Grain Rich Type-Chinese Famous Wine Jian Nanchun and Drinker Health. *Liquor-Making Science & Technology*, 2004, 31(4): 120–122 (in Chinese).

Index